Springer Aerospace Technology

The series explores the technology and the science related to the aircraft and spacecraft including concept, design, assembly, control and maintenance. The topics cover aircraft, missiles, space vehicles, aircraft engines and propulsion units. The volumes of the series present the fundamentals, the applications and the advances in all the fields related to aerospace engineering, including:

- structural analysis,
- aerodynamics,
- aeroelasticity,
- aeroacoustics,
- flight mechanics and dynamics
- orbital maneuvers,
- avionics,
- systems design,
- materials technology,
- launch technology,
- payload and satellite technology,
- space industry, medicine and biology.

The series' scope includes monographs, professional books, advanced textbooks, as well as selected contributions from specialized conferences and workshops.

The volumes of the series are single-blind peer-reviewed.

To submit a proposal or request further information, please contact: Mr. Pierpaolo Riva at pierpaolo.riva@springer.com (Europe and Americas) Mr. Mengchu Huang at mengchu.huang@springer.com (China)

The series is indexed in Scopus and Compendex

Gareth D. Padfield · Stefan van 't Hoff ·
Philipp Hofmeister · Linghai Lu · Mark White ·
Giuseppe Quaranta

Rotorcraft Certification by Simulation and Analysis

An Introduction to the Principles and Practices

Springer

Gareth D. Padfield
The School of Engineering
The University of Liverpool
Liverpool, UK

Philipp Hofmeister
German Aerospace Center
Braunschweig, Germany

Mark White
The School of Engineering
The University of Liverpool
Liverpool, UK

Stefan van 't Hoff
Royal Netherlands Aerospace Centre
Amsterdam, The Netherlands

Linghai Lu
Centre for Aeronautics
Cranfield University
Cranfield, UK

Giuseppe Quaranta
Department of Aerospace Science
and Technology
Politecnico di Milano
Milano, Italy

ISSN 1869-1730 ISSN 1869-1749 (electronic)
Springer Aerospace Technology
ISBN 978-3-031-86400-1 ISBN 978-3-031-86398-1 (eBook)
https://doi.org/10.1007/978-3-031-86398-1

This work was supported by Politecnico di Milano.

This Springer imprint is published by the registered company Springer Nature Switzerland AG
The registered company address is: Gewerbestrasse 11, 6330 Cham, Switzerland

Foreword

Aerospace is well renowned for the central role played by certification of products and the complexity associated to this endeavour. The continuous efforts to improve and harmonise certification requirements, methods and processes to show compliance have delivered over time undisputable results and have represented at the same time a reference for many other sectors.

Specifically, certification requirements have been the subject over the years of several discussions, focus groups, rulemaking committees, with more and more robust evidence to support the need for their amendments. They are so much metabolized by the industry that they are now part of the basic design requirements set of any new project, as no design organisation could afford to neglect them until it is too deep into the design and development process.

Not the same progress has been recorded around the means to demonstrate compliance to the rules. In fact, the role of testing and analysis have been very well established in such a way that compliance efforts and processes have been made more predictable and quantifiable; however, testing is by its nature limited and many times not conclusive. Analysis is very often necessary to interpret results and reach appropriate conclusions. Interpolations among test points are also appropriate and widely used, but the complexity inherent in rotorcraft has sometimes proved very challenging in spotting combinations of factors that could lead to critical conditions.

Moreover, there are many failure conditions or test cases that cannot simply be performed safely or with unambiguous results. It is also equally unpractical to verify the variety of production defects and combinations of tolerances by test.

A trained eye will also find numerous assumptions throughout the determination and execution of the compliance verifications versus certification requirements with past experience and standardisation playing a crucial role in the discussions between the applicant and the regulators. The whole process of compliance determination is heavy and lends itself to improvements.

A final consideration is on financial risk management. Some key complex flight tests can be set up and executed only when the associated program spending level is already very high, and the development team is running at full speed. Any set back then carries prohibitive consequences and in general brings expensive and suboptimal workarounds that ultimately will remain part of the end product. This has been the life of many project teams and is the source of continuous debates between engineering and financial departments in all organizations across the globe.

Therefore, investing in more extensive, more articulated and ambitious use of simulation has been seen as a silver lining by the industry, academia and regulators to cope with these issues but it comes with its own hurdles.

The need to define appropriate fidelity levels focused on specific characteristics and use cases is a key topic that is common between simulation and tests. Simulation must be handled with the same rigor of a physical test item to understand when good is good enough for the intended purpose.

The methodology and means to validate the outcomes of simulation and analysis come front and centre in the discussions to accept simulations not only as a development tool, but also as a suitable means of evidence.

This Book is therefore particularly welcome as it deals with all these aspects of the certification bringing a professional approach to the use of simulation and analysis together with their validation in combination with flight tests.

It puts things in the right perspective enabling project and certification teams to achieve more predictable and repetitive results at a more affordable cost thus representing a real competitive edge with the omen that more publications will stem from this one to push the boundaries even further.

I commend the hard work put together by the RoCS team with a big thank you for their endeavours.

Matteo Ragazzi
Head of Engineering and Innovation
Leonardo Helicopter Division
Milan, Italy

Preface

This Book presents guidance for the application of flight modelling and simulation and analysis[1] in support of certification for compliance with CS-27 and CS-29, SUBPART B—FLIGHT and other flight-related rules (e.g. CS-27/9, Appendix B, Airworthiness Criteria for Helicopter Instrument Flight). The guidance is presented in the form of a structured '*Rotorcraft Certification by Simulation*' (RCbS) process, starting from the relevant paragraphs, the Applicable Certification Rules (ACRs) in the Certification Specifications, through a comprehensive description of the process of capturing and assembling the flight simulation requirements, informed by judgements on Influence, Predictability and Credibility, and on into the detailed building of the three major elements of the process; the Flight Simulation Model (FSM), the Flight Simulator (FS), and the associated Flight Test Measurement System (FTMS).

A requirements-based approach is advocated and outlined, acknowledging the profound importance of assembling preliminary requirements, as complete as possible (Phase 1), before embarking on simulation development processes (Phase 2). 'Assembling' refers to the process of deriving requirements from the certification basis, the operational capability and engineering design data, considering the intended utility of the flight simulation. A detailed, and flexible, approach to requirements-capturing highlights the value in having multiple iterative/feedback routes from the build processes back to the requirements, ultimately to maximise the coherence between credibility and certification, and the requirements themselves. This dynamic is reflected in the content of this Book, where some details of the requirements might be uncovered and described in the development Phases 2a (FSM), 2b (FS) and 2c (FTMS).

In the current context, the power of M&S is contained within their ability to describe and predict flight behaviour. Used here, *describe* has the broad meaning that physical understandings can be gained for relationships between causes and effects. Such understandings are often blurred by the complexities of rotorcraft aeromechanics, and revealed only in limited ways by test data. The *predictive* capability of modelling is clearly critical in aircraft design and development and is expected

[1] Hereafter in this Book, 'analysis' is subsumed within 'modelling and simulation' activities.

to be key to both interpolation and extrapolation, and credibility assessment, in the RCbS process. Predictive fidelity is presented in this Book in terms of the three flight characteristics—trim, stability and response—that together provide a complete description of flight behaviour, including performance, controllability and manoeuvrability. A pilot's ability to engage with these characteristics determines the aircraft's handling qualities and connects with a companion requirement for flight simulators—perceptual fidelity. Good predictive and perceptual fidelity maximise the utilisation of RCbS. The guidance presented in this Book provides examples of metrics for quantifying the fidelity that is 'sufficient' for application to relevant ACRs. The concept of 'adaptable-fidelity' is introduced in this Book to emphasise that what might be sufficient is task-specific, or in our case ACR-specific. The concept of 'sufficient' fidelity is intended to reflect a technical quantification of the EASA term 'representative'.

A deep appreciation of the overlapping and interacting nature of predictive and perceptual fidelity is considered fundamental to the successful application of RCbS. Consequently, the importance of applicant experience and specialist technical skills in the effective use of the power of (rotorcraft) M&S is stressed; a message that should ring loud and clear throughout the Book.

Although there has been considerable progress over previous decades, this Book acknowledges that the status of rotorcraft M&S is far from perfect. Much remains to be done to enable full certification credit solely based on flight simulation. However, Industry and Certification Authorities have been working in this direction for some time, as reflected in CS27/29 para 21 (a), which states that proof of compliance with the rule must be met, '*by tests upon a rotorcraft of the type for which certification is requested, or by calculations based on, and equal in accuracy to, the results of testing*'. The guidance in this Book builds on this, adding substance to the meaning of '*equal in accuracy*'.

Specifically, we refer to the following four options:

(1) de-risking,
(2) critical-point analysis,
(3) partial credit, and,
(4) full credit,

when considering the Influence Levels for RCbS. Within each Influence Level, Predictability Levels are defined by the extent of interpolation and extrapolation between and beyond conditions (planned to be) validated by test data. A third dimension is added to the framework by consideration of the Confidence Level required for the application of RCbS. This 3-dimensional framework is reinforced by the manifold of domains within which RCbS is undertaken—the domain of physical reality (DoR), the domain of prediction (DoP), the domain of validation (DoV) and the domain of extrapolation (DoE). The latter refers to the extent of the DoP that is outside the DoV. It is within the DoE, particularly, that uncertainties and confidence levels in the modelling and simulation adopted need to be analysed and quantified. The confidence-ratio concept is introduced to reflect confidence in the prediction relative to the proximity to the applicable performance requirement. Such quantifications and judgements will inform the crucial 'Credibility Assessment' in Phase 3, that establishes whether the selected Influence Levels have been achieved for the ACRs.

As with all such endeavours in engineering, the process should commence with the production of an RCbS Project Management Plan (PMP), described as Phase 0 of the process. The PMP provides a framework for the whole RCbS process and is discussed further in Chap. 3, along with project documentation, data and configuration management and resource requirements. The first issue of the PMP will strictly be 'preliminary', noting that until the Requirements Specification for the selected ACRs is developed in preliminary form in Phase 1, and consolidated in Phase 2, what is achievable and the required resources can only be best estimates.

The importance of following the structured process is thus emphasised, so that steps are not missed and lessons learned from early adopters can be used in the continuous improvement of the process.

The guidance presented in this Book is intended to serve as a reference to support applicants gain an appreciation of a route to achieving RCbS. Furthermore, it provides a framework for community-wide debate and critical reviews ahead of any potential formal acceptance of such processes, and the publication of related standards, by the certification authorities.

The RoCS team recognise that establishing a comprehensive RCbS capability will require both short and long-term investments. Benefits in certification time and cost, safety and performance are likely to be accrued gradually, with initial applications rich in learning and capability-development experiences. The Book discusses possible routes towards developing such a RCbS capability, in terms of both technical breadth and depth, and drawing on certification cases (Chaps. 11–14) to exercise the process. Chapter 15 discusses possible 'next steps' along such routes, within the context of a long-term industrial strategy, addressing ACR options for early successes and capability development.

Finally, the Book makes use, in places, of the grammar modals should, shall, can, may, etc. Unlike in formal requirements specifications or rules, there is no intention to be prescriptive here or to differentiate between levels of modal importance. However, the RoCS team envisage the guidance laid out in this Book as a starting point in achieving a grander objective of defining a formal Acceptable Means of Compliance with Certification Specifications using M&S.

Liverpool, UK — Gareth D. Padfield
Amsterdam, The Netherlands — Stefan van 't Hoff
Braunschweig, Germany — Philipp Hofmeister
Cranfield, UK — Linghai Lu
Liverpool, UK — Mark White
Milan, Italy — Giuseppe Quaranta

Acknowledgements This Book was developed from one of the key outputs of the Clean-Sky2 project 'Rotorcraft Certification by Simulation (RoCS)—the Guidance Manual for Rotorcraft Certification by Simulation (RCbS)'. The authors would like to acknowledge support provided in a variety of ways from the following people and organisations throughout the project (2018–2024).

During the 2022 'public consultation' phase of the first release of the guidelines, the following provided insightful feedback:

Xavier Henriquel, Willem Doeland, EASA
John van Houdt, FAA
Riccardo Dona, EU JRC Ispra
Olaf Stroosma, TUDelft
Mike Jones, Systems Technology Inc.
Olivier Jeunehomme, Airbus

In the final stages of the production of this Book, the RoCS team sought advice on the approach to uncertainty quantification adopted; the team acknowledge the valuable feedback provided by:

Peta Hristov (Sofia University, formerly at the University of Liverpool) and Philipp Bekemeyer (DLR Braunschweig)

Throughout the RoCS project, support from the Associate Partners was critical to the successful outcomes, particularly the following individuals:

EASA	Francesco Paolucci, Hamdy Sallam (test pilot for the piloted simulations conducted in the Case Studies) and Mathilde Labatut.
Leonardo Helicopters	Andrea Ragazzi, Federico Del Grande, Fiorenzo Posterivo, Lorenzo Frigerio and test pilots Gianfranco Cito and Antonio Baldussi

The authors would like to express their gratitude to staff from their home organisations for support, particularly for analyses in the Case Studies and the preparation and conduct of the flight simulation trials:

Polimi	Matteo Daniele, Pierre Garbo and Andrea Zanoni.
NLR	Bart Timmerman, Richard Bakker, Paul Breed, Jur Creijnen, Guido Tillema, Maxim van Oldenbeek and Joris Fields.
DLR	Mike Jones, Pavle Scepanovic, Torsten Gerlach.
UoL	Chris Dadswell, Wajih Memon and Dylan Coyle; test pilots Andy Berryman, Christopher Brown and Mark Prior.

Finally, the RoCS team acknowledge the continual support throughout the project from staff at the Clean Sky 2 Joint Undertaking:

JU Vincent Foucart and Antonello Marino.

The authors are conscious that their Book is a better product because of the multiple reviews and feedback received, and consequent revisions made. However, the authors also acknowledge their ultimate responsibility for the end-product and continue to welcome feedback, including lessons learned, from the user-community as the RCbS process is put into practice.

Competing Interests The authors have no competing interests to declare that are relevant to the content of this manuscript.

About This Book

This Book presents the steps necessary for the application of rotorcraft flight modelling and simulation and analysis in support of certification for compliance with EASA CS-27 and CS-29, SUBPART B—FLIGHT and other flight-related rules (e.g. CS-27/9, Appendix B, Airworthiness Criteria for Helicopter Instrument Flight). The material is presented in the form of a structured 'Rotorcraft Certification by Simulation' process, starting from the relevant paragraphs in the Certification Specifications, through a comprehensive description of the assembly of flight simulation requirements, informed by judgements on Influence, Predictability and Credibility, and on into the detailed building of the three major elements of the process; the Flight Simulation Model, the Flight Simulator, and the associated Flight Test Measurement System. The latter feeds both the flight model and simulator development with real-world test data to support validation and fidelity assessment. The Book is intended to provide support to early adopters of simulation and analysis in the certification process, including those who have considerable experience and expertise in the use of Modelling and Simulation in support of design and development. It is acknowledged that there exists much good practice in the rotorcraft industry in this regard. However, what is presented herein is considered a significant step forward in the development of this practice, particularly in terms of the importance of a structured, requirements-based, process using adaptable-fidelity, descriptive and predictive simulation tools and associated pre-certification flight testing, to reach the goal of a credible flight simulation.

Contents

1 Introduction 1
1.1 Purpose and Scope 1
1.2 Background 2
1.3 Document Structure 3
References 6

2 Structure of the RCbS Process 9

3 Developing the RCbS Project Management Plan (Phase 0) 13
Reference 16

4 Requirements Capture and Build (Phase 1) 17
4.1 Introduction 17
4.2 Influence, Predictability and Credibility Levels 21
4.3 Flight Simulation Requirements 30
4.4 Summary: Phase 1 34
References 35

5 Rotorcraft Flight Modelling and Simulation 37
5.1 Introduction 37
5.2 Simulation Type 37
5.3 Strengths of Modelling and Simulation 39
References 40

6 Flight Simulation Model Development (Phase 2a) 43
6.1 Introduction 43
6.2 Flight Simulation Model Build 43
6.3 Verification and Validation 55
6.4 Summary: Phase 2a 76
References 77

7 Flight Simulator Development (Phase 2b) ... 79
7.1 Introduction ... 79
7.2 Flight Simulator Build ... 82
7.3 Operator Station ... 83
7.4 Environment System ... 84
7.5 Ground Reaction and Handling System ... 84
7.6 Crew Station Layout and Structure ... 85
7.7 Flight Controls and Forces ... 85
7.8 Visual Motion Cueing System ... 86
7.9 Sound Cueing System ... 87
7.10 Vestibular Motion Cueing System ... 88
7.11 Vibration Cueing Systems ... 89
7.12 Flight Simulator Verification ... 89
7.13 Flight Simulator Validation ... 91
7.14 Summary: Phase 2b ... 94
References ... 95

8 Flight Test Measurements for FSM/FS Development (Phase 2c) ... 97
8.1 Introduction ... 97
8.2 FTMS Design ... 98
8.3 FTMS Build ... 99
8.4 Calibration ... 100
8.5 Installation ... 100
8.6 Usage, Including Flight Testing ... 101
8.7 Summary: Phase 2c ... 103
Reference ... 103

9 Credibility Assessment and Certification Activity (Phase 3) ... 105
9.1 Introduction ... 105
9.2 Simulation Credibility and Uncertainty ... 107
9.3 Exploring DoV Fidelity Assessments Extrapolated into the DoE ... 116
9.4 Phase 3 Concluding Remarks ... 119
9.5 Summary: Phase 3 ... 119
References ... 119

10 Resourcing the RCbS Process ... 121

11 Guidance for Selected ACRs Within the Certification Specifications ... 125
11.1 Introduction ... 125
11.2 The HELIFLIGHT-R Flight Simulation Facility and Rating Scales Used in the RoCS Piloted Simulation Trials ... 128
References ... 136

12 Case Study 1: CS-27/29 Low Speed Controllability and Manoeuvrability ... 139
12.1 Introduction ... 139
12.2 Phase 1: Requirements Capture and Build ... 140
12.3 Phase 2a: FSM Build and Development ... 143
12.4 Phase 2b: FS Build and Development ... 151
12.5 Phase 3: Credibility Assessment and Compliance Demonstration ... 155
12.6 Concluding Remarks and Recommendations ... 159
References ... 165

13 Case Study 2: CS 29/27 Category A Confined Area Rejected Take-Off ... 167
13.1 Introduction ... 167
13.2 Phase 1: Requirements Capture and Build ... 168
13.3 FTM Development ... 169
13.4 Phase 2a: FSM Build and Development ... 169
13.5 Phase 2b: FS Build and Development ... 175
13.6 Piloted Simulation Assessment ... 180
13.7 Concluding Remarks and Recommendations from the RTO Case Study ... 189
References ... 193

14 Case Study 3: CS 27/29 Dynamic Stability Requirements ... 197
14.1 Introduction ... 197
14.2 Dynamic Stability as a Flying Quality ... 199
14.3 The LDO in the DoV and DoE ... 202
14.4 Piloted Simulation Assessment of the Impact of the LDO on Handling Qualities ... 212
14.5 Concluding Remarks and Recommendations from the Dynamic Stability Case Study ... 219
Appendix 1: Summary of Dynamic Stability Requirements in Nominal Conditions ... 221
Appendix 2: LDO Stability Characteristics (Damping and Frequency) ... 223
References ... 223

15 Finale and Potential Routes to the Adoption of RCbS ... 227
15.1 Introduction ... 227
15.2 Follow Process ... 227
15.3 Requirements Capture ... 228
15.4 Resource RCbS ... 228
15.5 Case Studies ... 229
15.6 What Next for Early Applications? ... 230

Abbreviations

A/C	Aircraft
ac	Aerodynamic Centre
AC	Advisory Circular (FAA)
ACR	Applicable Certification Rule
ADS-33	Aeronautical Design Standard-33
AFCS	Automatic Flight Control System
AlU	Aleatory Uncertainty
AMC	Acceptable Means of Compliance (EASA)
AR	Augmented Reality
ATC	Air Traffic Control
C&M	Controllability and Manoeuvrability
CFD	Computation Fluid Dynamics
cg	Centre of Gravity
CP	Certification Programme
CPA	Critical Point Analysis
CR	Confidence Ratio
CS	Certification Specification (EASA)
DO	Design Organisation
DoE	Domain of Extrapolation
DoFs	Degrees of Freedom
DoP	Domain of Prediction
DoR	Domain of Physical Reality
DoV	Domain of Validation
DS	Dynamic Stability
EASA	European Union Aviation Safety Agency
EP	Evaluation Pilot
EpU	Epistemic Uncertainty
ES	Environment System
FAA	Federal Aviation Administration
FAR	Federal Aviation Regulations
F-AW109	FLIGHTLAB model of AW109 Trekker

FCS	Flight Control System
FoR	Field of Regard
FoV	Field of View
FS	Flight Simulator
FSM	Flight Simulation Model
FT	Flight Test
FTG	Flight Test Guide
FTM	Flight Test Manoeuvre
FTMS	Flight Test Measurement System
HITL	Hardware-in-the-Loop
HQR	Handling Qualities Rating
HQs	Handling Qualities
HT	Horizontal Tail
IFR	Instrument Flight Rules
IGE	In Ground Effect
I-P	Influence-Predictability
LDO	Lateral-Directional-Oscillation
LHD	Leonardo Helicopter Division
LoI	Level of Involvement (of EASA) in a certification programme
M&S	Modelling and Simulation
MBDS	Multi-Body Dynamic System
MDA	Motion Drive Algorithm
ML	Machine Learning
MOC	Means of Compliance
MR	Mixed Reality
MTE	Mission Task Element
MUAD	Maximum Unnoticeable Added Dynamics
NPA	Notice of Proposed Amendment
OEM	Original Equipment Manufacturer
OMCT	Objective Motion Cueing Test
OS	Operator Station
OTW	Out-The-Window
PAO	Pilot Assisted Oscillations
PCE	Pace-Car Equivalent
p-model	Phenomenological Model
PMP	Project Management Plan
QTG	Qualification Test Guide
RCbS	Rotorcraft Certification by Simulation (process name)
RFM	Rotorcraft Flight Manual
RoCS	Rotorcraft Certification by Simulation (project name)
RPCs	Rotorcraft-Pilot Couplings
RTO	Rejected Take-Off
SAS	Stability Augmentation System
SCAS	Stability and Control Augmentation System
SME	Subject Matter Expert

STC	Supplemental Type Certificate
tc	Torsion Centre
TC	Type Certificate
TDP	Take-off Decision Point
TSR	Trim, Stability and Response formulation of flight behaviour
UAQ	Uncertainty Analysis and Quantification
UQ	Uncertainty Quantification
V&V	Verification and Validation
VeMCS	Vestibular Motion Cueing System
VFR	Visual Flight Rules
VR	Virtual Reality
VRS	Vortex Ring State
VzMCS	Visual Motion Cueing System
XWH	Cross-wind Hover

Definitions

Acceptable Means of Compliance Non-binding standards adopted by certification authority to illustrate means to establish compliance with ACR.

Accuracy The closeness of a parameter or a variable with the assumed true reference. It usually requires a metric to be quantified.

Aleatory Uncertainty The inherent variation associated with the physical system or the environment under consideration, e.g. the variation of geometric and or material properties due to manufacturing process. It is stochastic and irreducible below a certain threshold.

Applicable Certification Rule A rule, normally specified by a paragraph in a Certification Specification, that is considered a candidate (applicable) for compliance demonstration using modelling and simulation. ACRs include any Applicable Issue Paper, or Certification Review Item issued by the certification authority.

Average Pilot A pilot able to apply a normal level of skills required in the context of civil rotorcraft operations, upon completion of the appropriate flying training. An average pilot would not be expected to exercise exceptional piloting skills, alertness or strength in the course of their flying duties.

Comparison Error The difference between the result of an experiment, or any other referent, and the corresponding simulation result. It is indicated with the symbol δ_c. In this Guidance, the preferred terms are prediction error or validation error, the latter applicable in the domain of validation.

Compliance Demonstration The process of demonstrating that a product is compliant with defined rules or standards.

Component/super component A distinguishable element or collection of elements of a flight simulation model

Conceptual Model The collection of assumptions and abstractions applied to develop a physical model of the system of interest.

Confidence The degree of certainty (or uncertainty) an applicant has in the accuracy of the results from a test or simulation.

Credibility The qualitative and quantitative characteristics of a simulation that combine in being convincing or believable in its representation of flight behaviour for the ACR under investigation.

Credibility Assessment The process of quantifying the credibility of simulation results. In the DoV, the credibility can be supported by results from fidelity assessment. In the DoE, credibility is derived from the confidence the applicant or evaluator has in the extrapolation process and associated uncertainties.

Damping Ratio A characteristic measure describing how an oscillation in a system decays after a disturbance; for low damping, it is approximately proportional to how much the oscillation decays/grows in a cycle, as described in the logarithmic decrement method.

Data Pedigree A record of traceability of the data used to build the FSM. It should cover all aspects of data source, transmission, storage and processing to its final form used to build or validate the FSM.

Degrees of Freedom The set of independent variables that completely define the state of the flight simulation model.

Domain of Extrapolation The domain within which extrapolation of predictions are made to achieve certification at defined Influence Levels for an ACR.

Domain of Physical Reality The domain within which the laws of physics being used are adequately represented in the flight model and flight simulator. Since all models and simulations used in the RCbS process will include approximations to physical reality, this domain is strictly the region where the approximations are valid, reflecting the description 'adequately represented'.

Domain of Prediction The domain within which it is the intention to predict the behaviour of the aircraft or component and to use these predictions to achieve certification at the defined Influence Levels for an ACR.

Domain of Validation The domain within which test data will be used to validate the flight model or simulator and their components/features. Validation means a positive outcome has been achieved for the relevant metrics in the fidelity assessment. Within the DoV, interpolation is used to predict behaviour between validation points.

Domain of Verification The region of all conditions for which FSM/FS codes and implementations are deemed to be correct (i.e. function as intended) and solutions have been determined to possess the required accuracy.

Engineering Design Data The design data collected to develop a system or a product, including drawings, results of experiments performed during the product development, simulation models used during the design, previous experience on similar designs etc.

Epistemic Uncertainty The potential inaccuracy in any phase or activity of the modelling process that is due to a lack of knowledge or to intentional approximations applied by the analyst. It is potentially reducible by model improvements or by a better measurement technique employed to assess model parameters.

Experimental Error The difference between the experimental value, or the value of any other referent, and the true (unknown) value. It is indicated with the symbol δ_r.To this error, it is possible to associate an uncertainty u_r.

Fidelity Assessment The process of quantifying the fidelity of simulation results in the DoV.

Flight Simulation Flight simulation refers to either off-line desktop simulation, or real-time pilot-in-the-loop simulation in a suitable FS.

Flight Simulation Model A computational model of flight behaviour that can be created and analysed through the employment of software, to generate data useful to support the design, development and certification processes.

FSM Fidelity Fidelity of the FSM as reflected in the accuracy with which flight behaviour is modelled compared with the real aircraft or, more generally, the referent. It is assessed through the definition of one or more metrics to identify the FSM accuracy i.e. the predictive fidelity. The same FSM can have different fidelities, depending on the usage and on the prediction domain chosen. The concept of fidelity in this Guidance is generally associated with the ACR under investigation.

FSM Uncertainty Estimated variation in the results of the FSM due to factors inherent to the FSM and not to the referent used for validation.

Flight Simulator A device for enabling a pilot to fly tasks, e.g. associated with an ACR, in a virtual environment.

FS Fidelity How well the outputs of the flight simulator agree with the corresponding values in the referent (parameters that quantify the fidelity). The FS fidelity is composed by two parts: the predictive fidelity, that includes the FSM fidelity, and the perceptual fidelity, perceived by a pilot.

Handling Qualities As defined by Cooper-Harper in [1], "*Those qualities or characteristics of an aircraft that govern the ease and precision with which a pilot is able to perform the tasks required in support of an aircraft role.*"

Handling Qualities Ratings Ratings awarded according to the HQ rating scale developed by Cooper-Harper in [1].

Influence Level The extent to which the use of RCbS influences the Certification process classified in four levels—full credit, partial credit, critical point analysis, de-risking.

Input Error The error in a parameter used as input for the simulation model. It is indicated with the symbol δ_{inp}. To this error, it is possible to associate an uncertainty u_{inp}.

Mathematical Model Mathematical formulation of the relationships between cause and effect.

Metrics Normally a mathematical function to measure the distance between two elements: two points or results, or two sets of points or results.

Model Calibration The process of adjusting physical modelling parameters in a model to improve agreement with a referent (commonly used in other fields of application).

Model Component The subsystems or individual elements that make up the Flight Simulation Model. Typically, each FSM is composed of several components interconnected together. Collections of components might be referred to as a super-component.

Model Error Featuring in this Guidance as the FSM error, the error caused by the modelling assumptions or their numerical implementation. It is indicated with the symbol δ_{model} and is characterized by an interval rather than a unique value.

Model Tuning The process of adjusting model parameters to improve agreement with a referent. Can also be used in cases where the physics-based structure of the model is not considered critical.

Model Updating The process of updating model structure and content to improve agreement with a referent; the term more commonly used than model calibration in flight mechanics applications.

Modelling and Simulation Modelling and Simulation (M&S) is the use of a (conceptual, mathematical or numerical) model as a description or representation of a real system or phenomena for simulation by computational means. Modelling is the act of constructing a model; simulation is the execution of a model to obtain results.

Numerical Error The error due to the numerical algorithms employed to solve the problem. It is indicated with the symbol δ_{num}. To this error, it is possible to associate an uncertainty u_{num}.

Phenomenological Model A mathematical relationship between cause and effect that is created from measurements or high-order numerical methods (e.g. CFD). This term is used in the RCbS process with reference to, e.g., linear models derived from system identification, usually with a structure defined at the outset based on simpler, linear forms of the FSM. Other examples include models, such as derived from wind tunnel aerodynamic data, or models obtained by using Artificial Intelligence or Machine Learning algorithms. The latter algorithms present specific risks and challenges, particularly with regards to extrapolation, which are not addressed within this Guidance.

Physics-based A physics-based model is one where all relationships between cause and effect, inputs and outputs, are governed by the laws of physics. This is in contrast with phenomenological models, where relationships are generally constructed from measurements of cause and effect, inputs and outputs, possibly without regard for the underlying physical laws.

Physical Laws A scientific generalisation based on empirical observations of physical behaviour. Empirical observations are typically conclusions based on repeated scientific experiments over many years, and which have become accepted universally within the scientific community.

Pilot Assisted Oscillations A more general form of the MIL-HDBK-1797A [2] term 'pilot-induced oscillation'—sustained or uncontrollable oscillations resulting from efforts of the pilot to control the aircraft

Perceptual Fidelity Perceptual fidelity refers to the fidelity of the cues that are transferred from the simulator hardware to the pilot. In this Guidance, 'sufficient' fidelity refers to the case that the pilot is able to apply reactions that are close to those that will be experienced in flight. It is composed by many aspects and reflects the variety of sensorial inputs that can be acquired through the human body (visual, auditory, tactile and body movement perception).

Predictive Fidelity Predictive fidelity refers to the fidelity of the FSM or the FS, and their components/features, that can be quantified with defined metrics and tolerances.

Referent Data, information, knowledge, or experimental results against which a FSM or simulation can be compared. It can be real-world data, or results obtained using analogous systems or, in some cases, higher fidelity models.

Requirements The source description for how an entity (e.g. FSM, model component) should function, operate (including constraints) and interact with other entities through inputs and outputs. Requirements also apply to the results of fidelity analysis (e.g. in terms of metrics and tolerances) and uncertainty analysis (e.g. in terms of design data input and measurement precision). The associated requirements specification should be complete and traceable and testable within the V&V processes.

Risk The risk is the combination of the predicted severity of consequences and the likelihood, i.e., probability of occurrence. A risk can be reduced by addressing either of these two elements.

Rotorcraft Pilot Couplings Unintended excursions in a rotorcraft's attitude and flightpath caused by anomalous interactions between the pilot and the aircraft.

Sensitivity Analysis The study of how the variation of an output of, e.g. the FSM, can be attributed to different sources of variation in the model input and parameters.

States Variables required to completely define the degrees of freedom for describing flight behaviour. For example, a single degree of freedom mechanical system, whose dynamics is represented by a second order differential equation, requires two states to be modelled (often the position and the velocity of the degree of freedom). Rigid-body flight dynamics with six degrees of freedom of fuselage motion are described by nine states.

Simulation Error The difference between the simulation output and the true (unknown) value. It is indicated with the symbol δ_s.

Subject Matter Expert An individual having education, training and experience in a discipline (the subject matter), system and process.

Validation The part of the V&V process for determining the degree to which a model, or a simulation, is an accurate representation of the real world from the perspective of the intended uses of the model. It is conducted by comparing the model or simulation with a referent, extracted from the real word. The validation process aims to ensure that the FSM or FS meets the associated fidelity requirements.

Validation Pseudo-Standard An estimate of the standard deviation of the combination of the effects of input, numeric and experimental uncertainties.

Uncertainty It can be estimated in the domain of validation and must be extrapolated to the domain of extrapolation. Indicated with the symbol u_{val}.

Verification The part of the V&V process for determining that the FSM and FS accurately represent the underlying conceptual and mathematical models and their solutions. It is usually composed of two elements: code verification and solution verification. The verification process aims to ensure that any model meets the code structure and solution requirements, so functioning as intended.

Variables (Global) Variables associated with the FSM as a whole, at aircraft-level.

Variables (Local) Variables associated with FSM components.

Virtual Pilot A computer-pilot operating through defined algorithms to fly manoeuvres and tasks.

References

1. G. E. Cooper and R. P. Harper, (1969), The Use of Pilot Ratings in the Evaluation of Aircraft Handling Qualities, NASA TN D-5153, NASA.
2. Anon, Flying qualities of piloted aircraft, MIL-HDBK 1797, US Department of Defense Handbook, December 1997.

Symbols

p_1, p_2	Generalised flight envelope variables
u_{inp}	Input uncertainty
u_{model}	Model structure or form uncertainty
u_{num}	Numerical uncertainty
u_p	Prediction uncertainty
u_r	Experimental (referent measurement) uncertainty
u_{val}	Validation pseudo-standard uncertainty
u_{X_i}	Uncertainty associated with input X_i
x_i	Input in the context of input uncertainty analysis
M	Performance margin
N_r	Yaw damping derivative (1/sec)
R	Referent data, from experimental data (in the DoV), used to compute validation error
S	Result from Simulation prediction, used to compute validation error
U	Total uncertainty (combined from variety of sources)
V	Aircraft velocity
VNE	Never exceed speed
VY	Best rate of climb speed
α	Incidence
β	Sideslip
δ_3	Rotorblade flap-pitch coupling
δ_c	Comparison error
δ_{inp}	FSM input parameter error
δ_{model}	FSM model (structure/form) error
δ_{num}	Numerical error in FSM solutions
δ_p	FSM prediction error
δ_r	Experimental error in producing the referent data
δ_s	Simulation error

δ_{val}	Validation error ($S{-}R = \delta_p - \delta_r$)
ω, ω_n	Frequency and natural frequency of oscillation on the eigenchart
ζ, $\zeta\omega_n$	Relative damping and damping of oscillation on the eigenchart

Chapter 1
Introduction

1.1 Purpose and Scope

The purpose of this Book is to provide guidance on the multiple aspects to be considered when using flight simulation and analysis to support, either directly or indirectly (e.g. full-credit, de-risking), the demonstration of compliance with the flight-related rules within Certification Specifications CS-27/29 [1, 2]. The guidance is intended to facilitate the development and define the constraints for the effective use of flight simulation, to support, augment or replace flight testing in the demonstration of such compliance, without sacrificing the level of safety. The simulation may take the form of e.g. off-line, desktop simulations using a stand-alone Flight Simulation Model (FSM), or of real-time piloted simulations in a suitable Flight Simulator (FS).

The scope of the material in this Book is broad, encompassing a requirements-based approach to the development, verification and validation (V&V), and usage of flight simulation models and associated flight simulators. Through modelling and simulation (M&S), rotorcraft flight mechanics is described and quantified, and linked with the Certification Specifications, within the behavioural elements of trim, stability and response and associated flight handling qualities. Fidelity assessment through simulation validation is a major aspect of the approach described in this Book, hence attention is also given to the requirements for test data and the development of the Flight Test Measurement System (FTMS).

Following good practice in design and manufacturing, the guidance is assembled in the form of a structured process, proposed to be followed by applicants to ensure maximum benefit from the adoption of flight simulation as an alternative means, or to otherwise support the showing of compliance with the applicable rules.

The content has taken into consideration the outputs from various related activities including the European Union Aviation Safety Agency (EASA) Proposed Certification Memoranda CM-S-014 Issue 01 on M&S for CS-25 Structural Certification Specifications [3], and the parallel evolution of the Proposed Means of Compliance (MOC) with the Special Condition VTOL (MOC SC-VTOL) [4]. The application of

G.D. Padfield et al., *Rotorcraft Certification by Simulation and Analysis*, Springer Aerospace Technology, https://doi.org/10.1007/978-3-031-86398-1_1

Rotorcraft Certification by Simulation (RCbS[1]) to the certification of eVTOL aircraft was not contained within the terms and conditions of the RoCS project. However, over the duration of the 'project', as the content of EASA's SC-VTOL has evolved, it seems clear that the read-across to this application is a natural extension. It could be argued that RCbS is highly relevant to the certification of the advanced flight control technologies and handling qualities of eVTOL aircraft. While this Book does not explicitly refer to the rules in the SC-VTOL and associated means-of-compliance, we consider that a common framework is entirely appropriate as a starting point for the adoption of RCbS for certification of these novel technologies.

1.2 Background

To quote from the specifications, proof of compliance with CS-27/29 Subpart B must be obtained by *"tests upon a rotorcraft of the type for which certification is requested, or by calculations based on, and equal in accuracy to, the results of testing."* As in the Federal Aviation Administration (FAA) Advisory Circular AC-29.21(a) [5], the term "*calculation*" includes flight simulation. The term "*equal in accuracy*" is subject to interpretation and will be addressed in this Book in the material on fidelity and credibility assessment and measured in terms of fidelity and uncertainty metrics defined by the applicant and agreed by the certification authority (in Phase 1, see Sect. 1.3).

Flight testing is costly, time consuming and may carry with it significant risk. It is anticipated that certification compliance demonstration through flight simulation, under the right conditions, may yield benefits in all these three aspects. However, to deliver such benefits, a concerted effort is required on the part of the applicant to develop, validate and maintain a simulation environment that is of sufficient fidelity for the application and is exercised within the limits of its validity. Outside the limits of proven validity (outside the domain of validation (DoV)), in the Domain of Extrapolation (DoE), predictions might be used to enable flight simulation to reach areas of the domain of prediction (DoP) that, for various reasons, are not populated with test data, e.g. Applicable Certification Rules (ACRs) associated with high-risk failure conditions, or areas of the envelope that require relocation to high-altitude test sites. The Book expands on the important concept of 'sufficiency', and the various 'domains' in which M&S are used, in Chapters 4, 6 and 7.

EASA's Part 21 [6] defines the general principles under which flight simulation may be proposed as an acceptable alternative to flight testing. Similarly, in the FAA's AC 25-7D §3.1.2.6 [7], simulation is taken as one of the elements, or possibly in some cases as the only element, to inform decision-making on airworthiness. Paramount to

[1] The term Rotorcraft Certification by Simulation, RCbS, is applied to the process developed in the RoCS project, to distinguish from the project name. It is recognised that Certification by Analysis (CbA) is also commonly used to mean, essentially, the same thing. Throughout this Book the RCbS process is intended to include analysis.

the acceptance of a simulation approach for certification purposes, is that it must be shown that the simulation leads to credible predictions of flight behaviour. Conventionally, the prediction error is determined by comparisons between (ground and/or flight) test data and analytical/numerical results, performing a set of analyses that fall under the term 'Validation'. Beyond validation, for the usage of simulation to support airworthiness decision-making, it is necessary to show that the models are also 'Credible', in that the uncertainties in the predicted outcome, beyond and within the validation domain, are quantified and acceptable. Validation is addressed within the relevant Chapters 4–8, while credibility is introduced in Chapter 4 and addressed in more detail in Chapters 6 and 9. The topic of uncertainty quantification (UQ) is threaded throughout the Book, acknowledging the importance of reducing uncertainty to the minimum possible; uncertainty in the measurements, the engineering data inputs, the FSM form, the solution processes and so on. While this importance is highlighted, the RoCS project did not have access to specific uncertainty data related to the Case Studies used to exercise the RCbS process, hence the coverage is somewhat limited, drawing on some of the published practices to illustrate UQ theory.

The idea of using simulation and analysis for certification is not peculiar to the aerospace sector, and other technological sectors are pursuing a similar path. It is worth noting the specification for the type of approval of the automated driving system of fully automated vehicles adopted by the European Parliament, where in Part 4 the principle of credibility assessment of models for certification are laid down [8].

While the guidance developed in this Book is intended to be equally applicable to Original Equipment Manufacturer (OEM) Type Certificate (TC) and Supplemental Type Certificate (STC) applicants, it is recognised that a lack of access to OEM engineering design data and development flight test data, such as might be the case for STC applicants, may skew the cost–benefit trade analysis in favour of flight testing. Equally, it is understood that applicants may elect to exploit an existing FS and/or FTMS, provided the requirements specified and agreed in Phase 1 are satisfied.

As the state-of-the-art in flight modelling and simulation is continuously evolving, it is expected that their utility and application for certification purposes will increase over time. This Book attempts to provide a route for such increased, more extensive, application. Furthermore, ground testing and/or pre-certification, developmental flight testing, for the (sole or partial) purpose of validation, are expected to remain an integral part of ensuring simulation fidelity and credibility. As such, the requirements for pre-certification testing become part of the process described in this Book.

1.3 Document Structure

The material in this Book falls under two main categories. The first category (Chapters 2–9) contains the description of the overall Rotorcraft Certification-by-Simulation (RCbS) process, commencing with an overview of the process

(Chapter 2). Chapter 3 is concerned with the project management plan (PMP) (Phase 0), and how the PMP needs to be harmonised with an applicant's Certification Plan. Chapter 4 describes the flight simulation requirements capture and build process (Phase 1). Then, Chapter 5 introduces the types of flight simulation that might be used, followed by Chapters 6 and 7 addressing simulation model and simulator-building, verification and validation (V&V) and fidelity assessment (Phases 2a and 2b), with Chapter 8 addressing the same for the FTMS development (Phase 2c). Credibility assessment and certification are presented in Chapter 9 (Phase 3).

Within this first category (Chapters 2–9), the following activities and sub-processes are featured; V&V of the FSM and FS, calibration of the FTMS, FS fidelity assessment, model-updating and credibility assessment.

The second category of guidance material is contained in Chapters 10–15. Chapter 10 discusses briefly the topic of Resourcing the RCbS process, emphasising the importance of early investment to create a robust virtual engineering framework.

Chapter 11 summarises the guidance for specific ACRs, the so-called Case-Studies, within the Certification Specifications, drawn from the results of assessments with state-of-the art FSMs and FSs. The opportunity is taken to illustrate aspects of the RCbS process that were exercised in these Case Studies, and described in detail in Chapters 12–14. Concluding remarks are drawn together in Chapter 15, along with suggestions for next steps along the routes forward for the early adopters of the RCbS process outlined in this Book.

The content of these Chapters is summarised in Fig. 1.1.

In extended summary form, the RCbS Phases are as follows:

(a) Phase 0; RCbS Project Management Plan (PMP), addressing,

    i. resources and timescales,
    ii. dependencies and constraints,
    iii. risks and mitigations,
    iv. process control, documentation, configuration and data management,
    v. structure for documenting the RCbS certification case,
    vi. Output; the RCbS PMP used to provide governance for all activities in Phases 1–3.

(b) Phase 1; assembly of the (preliminary) RCbS Requirements Specification including,

    i. ACR(s) from the Certification Specifications are identified for RCbS,
    ii. the four domains within which the RCbS will be carried out are defined (DoV, DoP, DoE, Domain of Physical Reality (DoR)),
    iii. the Influence and Predictability Level matrices are defined for the selected ACRs,
    iv. the relevant aircraft design data are collected together with related uncertainties,
    v. preliminary description of expected complexity content for the FSM, FS and FTMS needed to achieve 'sufficient fidelity' for each of the selected ACRs,

- vi. analysis and metrics for DoV fidelity assessment, together with tolerances for sufficiency, are defined, in preparation for meetings with certification authorities,
- vii. definition of test data requirements to characterise the domain of validation including programme for pre-certification flight trials and ground tests,
- viii. analysis and metrics for uncertainty quantification and credibility assessment are defined,
- ix. Output; the (preliminary) RCbS Requirements Specification assembled based on the above, using a comprehensive descriptive framework.

(c) Phase 2; development of the FSM, FS and FTMS based on the (preliminary) Requirements Specification

- i. FSM build, V&V, and fidelity assessment, including updating/tuning,
- ii. prototype FSM supplied to the FS and FTMS developers to support parallel activities,
- iii. FS build, V&V, and fidelity assessment, noting that legacy facilities could be used which may, or may not, require modification,
- iv. FTMS build, V&V, and fidelity assessment, noting that legacy facilities and approaches could be used which may, or may not, require modification,
- v. conduct pre-certification ground and flight test programme to support validation of FSM and FS,
- vi. multiple iterative pathways managed and exercised as required throughout Phase 2,
- vii. uncertainty quantification undertaken throughout the domain of prediction,
- viii. updates to Requirements Specification based on Phase 2 activities,
- ix. Outputs; FSM and FS verified and validated for intended purpose, included in fidelity and uncertainty assessment reports.

(d) Phase 3; Credibility assessment and Certification

- i. Credibility assessments undertaken based on the results from Phase 2 fidelity and uncertainty analyses,
- ii. results assembled and presented to certification authorities to make the ‘means of compliance’ case in the certification,
- iii. based on the feedback from Phases 2 and 3, the Requirements Specification is updated to constitute a formal element of the case for RCbS for the selected ACRs,
- iv. RCbS certification tests performed for relevant ACRs.
- v. Outputs; updated RCbS Requirements Specification and Type/Supplemental-Type Certificate documentation.

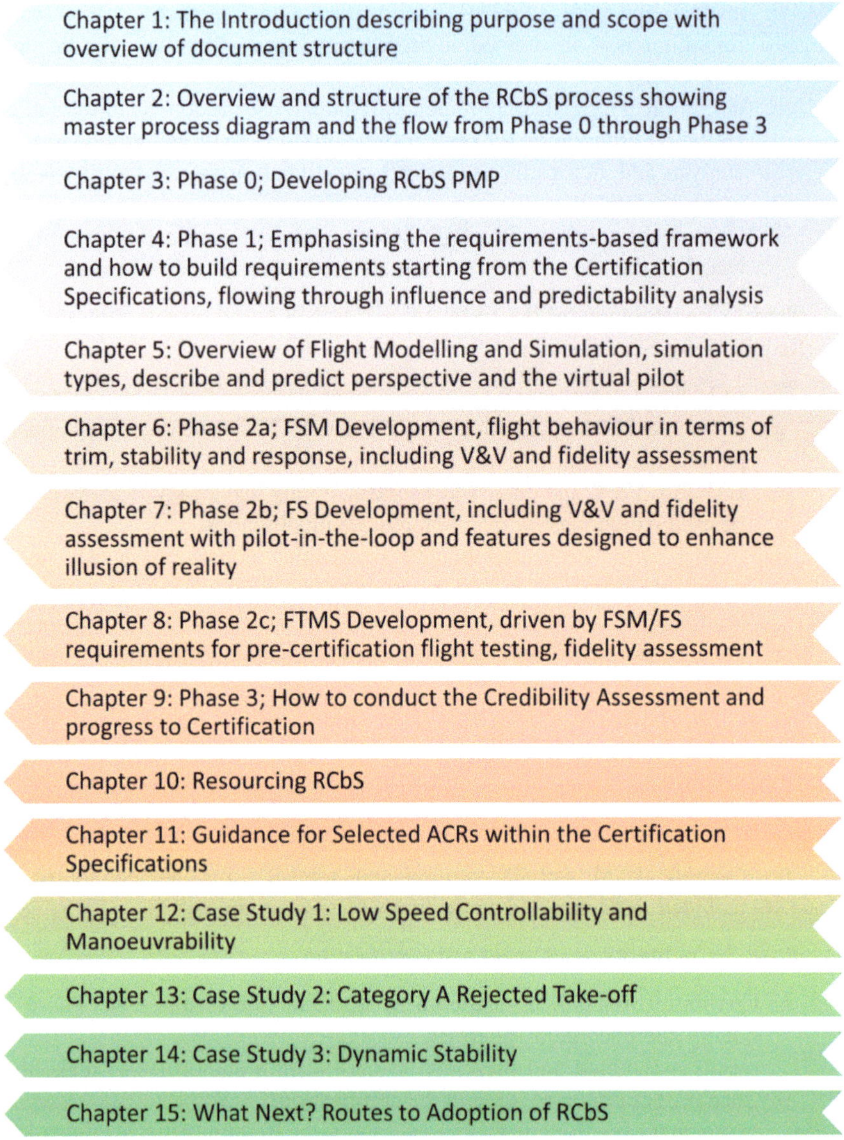

Fig. 1.1 RCbS process in brief

References

1. anon (2021) CS-27 Certification specifications, acceptable means of compliance for small rotorcraft. EASA
2. anon (2023). CS-29 Certification specifications, acceptable means of compliance for large rotorcraft. EASA

3. anon (2020) Notification of a proposal to issue a certification memorandum: Modelling & simulation—CS-25 structural certification specification. EASA
4. anon (2022) Third publication of proposed means of compliance with the special condition VTOL. EASA
5. anon (2014) AC 29–2C Certification of transport category rotorcraft. Federal Aviation Administration
6. anon (2022). Easy access rules for airworthiness and environmental certification (Regulation (EU) No 748/2012) EASA
7. anon (2018) AC 25–7D Flight test guide for certification of transport category airplanes. Federal Aviation Administration
8. European Commission (2022) Automated cars-technical specifications. Retrieved July 2022, from https://ec.europa.eu/info/law/better-regulation/have-your-say/initiatives/12152-Automated-cars-technical-specifications_en.

Chapter 2
Structure of the RCbS Process

The comprehensive and structured RCbS process presented in this Book is illustrated in Fig. 2.1, with activities in each 'box' having dedicated Chapters, Sections or sub-Sections. Following on from the creation of an RCbS PMP in Phase 0, the RCbS process is organised in three main subsequent, but iterative, Phases. To recap, the four phases are.

The comprehensive and structured RCbS process presented in this Book is illustrated in Fig. 2.1, with activities in each 'box' having dedicated Chapters, Sections or sub-Sections. Following on from the creation of an RCbS PMP in Phase 0, the RCbS process is organised in three main subsequent, but iterative, Phases. To recap, the 4 phases are:

- Phase 0: Developing the RCbS PMP,
- Phase 1: Requirements-capture and build,
- Phase 2: FSM development (2a), FS development (2b) and FTMS development (2c),
- Phase 3: Credibility assessment and Certification activities.

In the above, and throughout this Book, we assume that the applicant is using piloted simulation in the RCbS process. This need not be the case and, in such situations, the certification credits may only be sought from the use of the FSM in the DoV and DoE. In this case Phase 2c will still be required, of course.

It is emphasised that the phases are to be managed to enable the multiple iterative cycles highlighted, to ensure that the results of any assessment (e.g. Verification, Fidelity, Credibility) can take the applicant back to a previous phase or sub-phase, as required. The Applicable Certification Rules (ACRs) themselves are input to the

G.D. Padfield et al., *Rotorcraft Certification by Simulation and Analysis*, Springer Aerospace Technology, https://doi.org/10.1007/978-3-031-86398-1_2

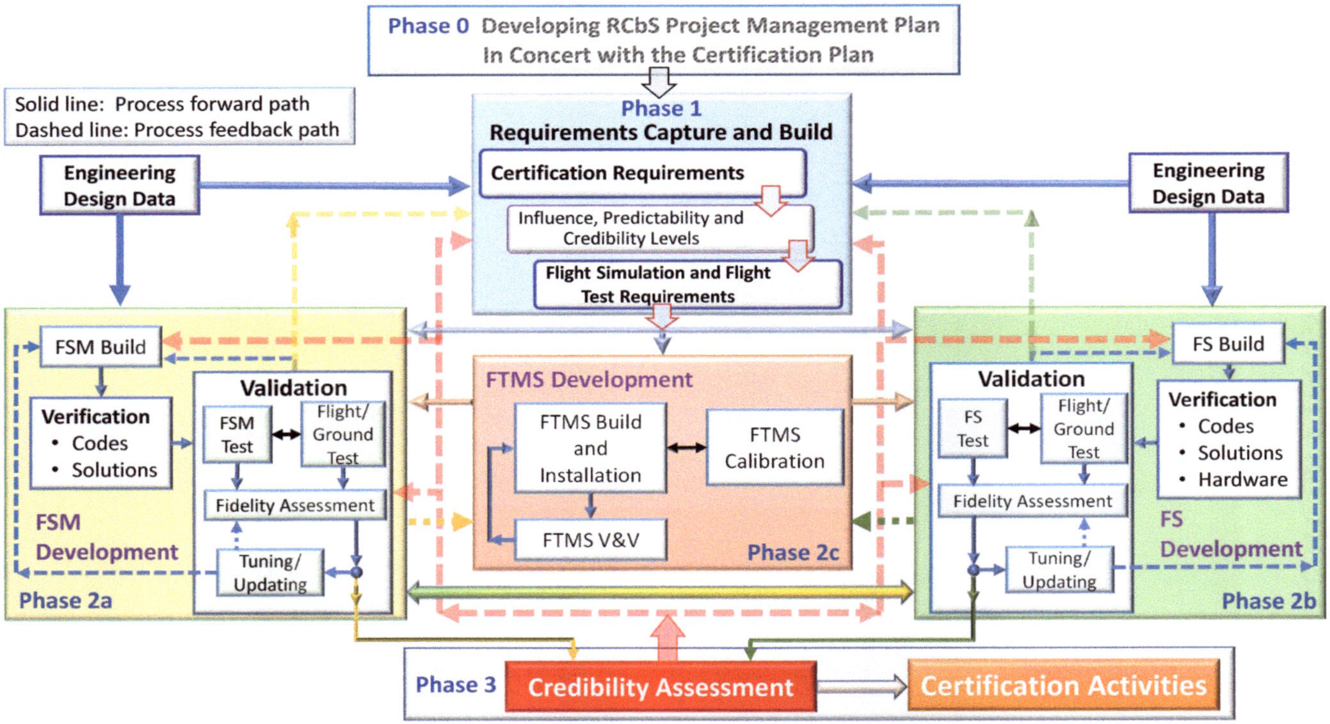

Fig. 2.1 Overall structure of the certification by simulation process. [FSM (Flight Simulation Model), FTMS (Flight Test Measurement System, FS (Flight Simulator)]

'Influence/Predictability/Credibility Levels' activity, which act as input to assembling the FSM and FS Requirements—the drivers for the whole process. These requirements are also informed by inputs from the Design Organisation's (DO's) engineering data (e.g. geometry, structure, inertial properties, component performance), including input parameter uncertainties. The process diagram in Fig. 2.1 uses solid lines to describe forward progress through the process. Recognising that activities in a phase or sub-phase might need to be updated based on results from a susequent phase, particularly fidelity and credibility assessment, dashed lines are used to highlight the return paths for iterative cycles. The ACRs and Engineering Design Data that feed Phase 1 are, of course, pre-defined inputs to the RCbS process. In this context, the ACRs include the associated Acceptable Means of Compliance (AMC), as well as any applicable Issue Paper, or Certification Review Item issued by the certification authority. It is particularly important in the RCbS process that the engineering data include comprehensive references for the data sources and any quantified uncertainties. The latter will be important for the uncertainty analysis and quantification that supports validation and credibility assessment.

It is recommended that, in the adoption of this RCbS process, progress from one phase of the process to the next be managed by reaching consensus between the applicant and the authority. This is particularly important for the requirements capture phase and the planning of the simulation and flight test campaigns, but also for decision-making related to fidelity and credibility assessment, e.g. metrics and tolerances.

To emphasise, the approach in this Book is driven by a requirements-based doctrine. It is well understood that the way requirements are expressed and quantified can evolve with application; they need testing to assess their veracity. We use the terms requirements 'capture and build' to emphasise the creative process involved here. Hence, iterative cycles are used extensively to allow all sub-processes to be improved based on the results of their application. Note that the requirements for the FSM, FS and FTMS may vary between ACRs, suggesting a tailored simulation development for a given application, highlighting the adaptable-fidelity concept.

Chapter 3
Developing the RCbS Project Management Plan (Phase 0)

This Chapter addresses what might be described as the administrative aspects of the RCbS process and application. Sometimes, and mistakenly, considered as 'second order' to the main development and creative activities, these aspects are, in this Guidance, put forward as equally critical to achieving success. Although there are currently no formal standards for the qualification of the simulation tools and data of the RCbS process, this situation may change as the experience builds. In such a scenario, methods, data management, standards and practices adopted must be fully documented and agreed with certification authorities to allow simulation to be used as a MOC. This Guidance therefore recommends a comprehensive approach to developing the PMP, the controlled development of the FSM, FS and FTMS and the associated configuration/data management; described as Phase 0 of the RCbS process.

Of course, the exercise of a management system already features as part of a certification activity. Applicants normally initiate a Certification Plan (CP) with EASA early in the development cycle (often well before the first flight) to establish the certification basis, the compliance checklist and the Level of Involvement (LoI) of EASA within the framework described in EASA Part 21 Easy Access Rules[1] [1] Therefore, Phase 0 of the RCbS process, as described in this Book, should be seen in the context of the Applicant's overall CP, described in AMC 21.A.15(b) and AMC 21.A.93 (b) [1]. The Guidance herein is not formulated specifically in the 'language' of the CP, but adopters should recognise that the material recommended needs to be welded into the CP in an appropriate manner to ensure consistency and avoid duplication.

As highlighted in the Introduction, the Phase 0 should address, for each selected ACR,

[1] Part 21 Subpart B, paragraph 21.A.15, "Application", describes the case for Type Certificate and Restricted Type Certificate while paragraph 21.A.93, "Application", describes the case for changes to Type Certificate and Restricted Type Certificate.

G.D. Padfield et al., *Rotorcraft Certification by Simulation and Analysis*, Springer Aerospace Technology, https://doi.org/10.1007/978-3-031-86398-1_3

i. Resources and timescales,
ii. Dependencies and constraints,
iii. Risks and mitigations,
iv. Process control, documentation, configuration and data management,
v. Preparation for documenting the RCbS certification case.

The first three numbered items are standard in project management; as is the fourth, but because of the potential evolution of requirements and configurations in RCbS, the fourth has added importance. It should also satisfy the dual purpose of providing comprehensive documentation of the applicant's RCbS process, and how it is managed, for the benefit of both the applicant and the certification authority. The process control and material management will underpin any required 'qualification' of FSMs and FSs for use in RCbS. Applicants should therefore formalise a controlled development and configuration/data management process, mirroring the phases described in this Book. This includes the systematic documentation of all relevant information necessary to enable the authority to understand the methodologies used, the underlying assumptions and limitations involved in Phase 2 developments, and to assess the validity of the simulation results and consequent Credibility assessments in Phase 3.

All these elements are clearly listed as part of the information that is typically expected by EASA whenever applying for certification (see AMC 21.A.15(b) and AMC 21.A.93 (b) [1]). In particular, the certification programme described in AMC 21.A.15(b) requires the applicant to provide "*a compliance checklist that includes a breakdown of the demonstration programme including a proposal for the means of compliance and the related compliance documents*". Additionally, "*when the compliance demonstration involves analyses/calculations, a description/identification of the tools (e.g. name and version/release of the software programs) and methods used, the associated assumptions, limitations and/or conditions, as well as of the intended use and purpose*" should be provided. It is also important to have a project schedule, including milestones. All items in the previous list should be included in the compliance checklist to allow the certification authority to determine its expected LoI.

The emphasis on creating a requirements-based framework for developing and validating, in parallel with pre-certification flight testing (Chapter 8), facilitates such a formalised approach. The requirements for the FSM and FS and their V&V, as well as a documented narrative on how the requirements have been met, or not, must be captured in a configuration/data management process.

The formal practice of (FSM/FS) configuration/data management facilitates appropriate representations with respect to the expected certification configuration(s), traceability of the results, and repeatability for future analyses and tests. The requirement for configuration/data management extends to the simulator hardware, even if generic subsystems are used (e.g. reconfigurable control loading system). Deviations from the expected certification configuration should be documented and justified. Records of the relevant information and data shall be retained in accordance with, e.g. Part 21.A.5 requirements [1]. In addition, it is suggested that a version control system is used to keep track of the changes applied to models, recording all changes applied, and tracking the complete history of changes through a repository that contains the historical evolution of the FSM or FS. Several tools exist that can perform this task automatically, allowing the user to be able to retrieve the whole history of a model.

The documentation effort also extends into compliance demonstration. In this context, AMC 21.A.20(c) contains the specification for producing such compliance documentation: to quote, "*Compliance documentation comprises one or more test or inspection programmes/plans, reports, drawings, design data, specifications, calculations, analyses,* etc.*, and provides a record of the means by which compliance with the applicable type-certification basis, the operational suitability certification basis and environmental protection requirements is demonstrated*" [1].

Items to be addressed in the RCbS configuration/data management and compliance documentation include:

(a) FSM/FS requirements specification, including how different FSM/FS variants are to be used and relate to one another,
(b) Data structures and related sources and uncertainties,
(c) V&V and UQ processes and results, including model tuning/updating,
(d) Definition and rationale/justification for the four domains (DoV, DoP, DoE, DoR),
(e) Problems relating to FSM and FS, e.g. configuration data and physics modelling,
(f) Interpolation and extrapolation,
(g) Experience and expertise being applied to RCbS by the applicant,
(h) Documentation and record-keeping processes.

Known problems, as referred to under e), would typically include deficiencies, process deviations and errors in the definition or implementation of the FS and FSM. These problems and their impact and/or mitigation should be documented and communicated in dedicated Problem Reports.

As described in Chapters 6 and 7, it is acknowledged that several versions of the FSM and FS are likely to be used in the RCbS process, addressing different Influence-Predictability combinations. It is recommended that a common framework for the different variants is used, that forms a core in the configuration management. The documentation of the V&V and UQ processes shall include details on the relevance and robustness of the selected metrics, and the model-updating methods that have been or will be applied. The overall documentation shall also include a description of the FSM and the simulator hardware. As detailed in Chapters 6 and 7, the components of the model and simulator can be described in terms of the requirements that the component or feature is serving, addressing functions, modes of operation, data structures, inputs and outputs, constraints and interfaces with other components.

The configuration and data management documentation covers a wide range of topics, in many ways mirroring the structure of this Book. The early adopters of the use of M&S in support of certification will have the opportunity to shape the development of this approach, identify the critical issues, and highlighting the strengths and weaknesses of different methods.

Reference

1. anon (2022) Easy access rules for airworthiness and environmental certification (Regulation (EU) No 748/2012). EASA

Chapter 4
Requirements Capture and Build (Phase 1)

4.1 Introduction

Before commencing the development of the RCbS process, it is necessary to understand the problem scope and determine the objectives in terms of desired outcomes and required accuracy. These understandings and determinations have both a specific perspective, i.e., related to a specific ACR, and a general perspective, i.e. related to general flight behaviour throughout the flight envelope. The understandings and determinations are captured within a set of requirements that the FSM, the FS and the FTMS must satisfy. In other words, RCbS is a requirements-based process as illustrated in Fig. 4.1 extracted from Fig. 2.1 The descriptive verbs 'capture' and 'build' are used here to emphasise the constructive nature of assembling requirements. There is a parallel here with capturing requirements during the preliminary design of a rotorcraft, where the requirements firm up as trade-off analyses are conducted on the design parameters. And, as with design trades, there are essential (fidelity) requirements, regarded as sufficient for application to certification. But, what do we mean by sufficient?

Introduction to Fidelity Sufficiency. The requirements-capture phase is intended to ensure that the (complexity) content within the FSM, the FS and the FTMS is appropriate to achieve this sufficiency. In addition to the specific and general perspectives described above, the concept of sufficient fidelity has another two dimensions; a predictive dimension, quantified by metrics and associated tolerances, and a perceived dimension, where an evaluation pilot (EP) provides a fidelity assessment of the FS to be used in the RCbS process. The EP's subjective fidelity assessment can also be supported through quantitative means such as by analysis of control activity (adaptation) and (comparable) task performance, as well as the recording of the visual scan patterns. As noted above, for the FSM, the acceptable differences between simulation and flight are quantified in terms of tolerances for the agreed metrics. Such tolerances will be ACR-specific and may evolve throughout Phases 1

G.D. Padfield et al., *Rotorcraft Certification by Simulation and Analysis*, Springer Aerospace Technology, https://doi.org/10.1007/978-3-031-86398-1_4

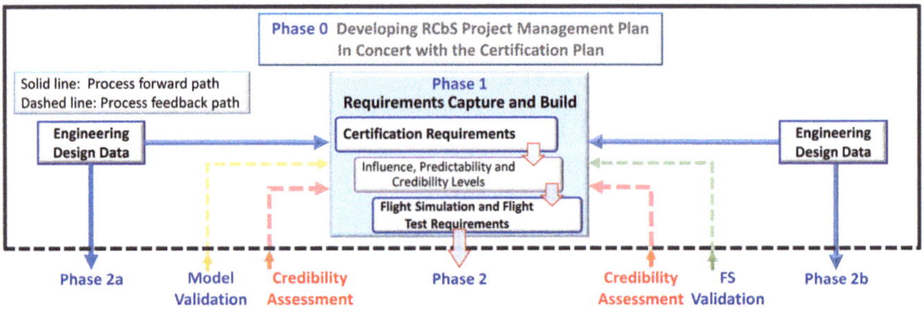

Fig. 4.1 Requirement capture and build phase of the RCbS process

to 3, e.g. when predictions are close to a performance or control margin limit, when the acceptable tolerances may need to be lowered.

However, Phase 1 is where the applicant proposes what metrics and tolerances are to be used for FSM/FS fidelity assessments in Phase 2, and what uncertainty metrics and tolerances are specified for credibility assessments in Phase 3. Within the trim-stability-response (TSR) characterisation of flight behaviour, example fidelity metrics include [1]:

(a) Trim; control positions/margins, pitch/roll attitude angles, fuselage incidence/sideslip for varying vertical/horizontal flight paths, static stability parameters,
(b) Dynamic stability; relative damping and natural frequency of oscillatory modes, time to half/double amplitude for aperiodic modes,
(c) Response; control power (maximum response of associated response type, e.g. deg/s for rate command response type, deg for attitude response type), on and off-axis response quickness (ratio of peak rate to attitude change in a discrete attitude change manoeuvre), attitude bandwidth and phase delay, cross-couplings (e.g. yaw from collective).

The stability and response metrics cited above are commonly used in classical flight mechanics to judge the quality of a simulation, but are often restricted to small excursions from a trim condition, excited by single axis control inputs. A legitimate question is then how relevant are these classical metrics to flight test manoeuvres (FTMs) that involve moderate or large departures from a trim condition, with multiple axes closed-loop control? Examples are the FTMs exercised in the Case Studies reported in Chaps. 11–14. For the Category A, confined-area, rejected take-off (RTO), flight behaviour (trim, stability) in hover, climb and descent are relevant, as well as transient behaviour (response) following the engine failure and during the final cushion and touchdown. To strengthen the credibility of the fidelity assessments derived from classical metrics, virtual pilot models can be used to 'drive' the FSM with flight data (e.g. controls or flightpath) to facilitate comparisons of the unconstrained dynamics. Chapter 13 presents example results for the RTO manoeuvre. Alternatively, longer term, integrated error cost-functions (e.g. NATO AVT-296 [2]) might be used to quantify fidelity.

In terms of tolerances, one approach is to define an acceptable % difference between flight and simulation (e.g. 10% of the flight measurement of control position, or attitude rate during a defined response time); another might be to define an absolute value (e.g. 0.5° attitude, 0.05 relative damping). It is easy to imagine a compendium of results of fidelity assessment generated in Phase 2, as evidence for the certification case in Phase 3. For each flight condition, there would be 'green' points where the tolerances (for trim, stability and response) have all been met, or 'amber' points where defined excursions outside the tolerances are perhaps allowed. Red cases could indicate cases where some tolerances are outside the sufficiency standards, highlighting the need for further FSM development or, ultimately, certification flight testing. The concept of adaptable-fidelity is advocated, where the subsets of fidelity metrics essential for sufficiency in the selected ACRs are defined in Phase 1, with associated justification, for presentation to the authority. The topic of required fidelity is returned to in Phase 2a, detailed in Chap. 6.

The Starting Point in Phase 1. To recap, the Requirements-capture and build phase starts with the identification of the ACRs, drawn from the CSs and associated material, for which simulation is foreseen to play a role in the compliance demonstration. A second key input to Phase 1, shown in Fig. 4.1, are the engineering design data that will be required to assemble the FSM and FS. Such data should be documented with their uncertainties (e.g. uncertainty in fuselage moments of inertia might be ± 5% of a nominal value based on product variability or measurement scatter). For reference, in the traditional 'certification by flight test' process, the following elements would be defined for the test campaign related to a specific ACR, or set of ACRs dealing with the same domain:

(a) Flight envelope to be investigated,
(b) Aircraft configurations to be tested, defined in terms of weight and balance (c.g. position),
(c) Flight Test Points (airspeed, altitude etc.)
(d) Flight Testing techniques (typically these are included in the AMCs),
(e) Flight Test Instrumentation required and associated minimum set of parameters to be recorded,
(f) Qualitative crew assessments,
(g) Data post-processing required,
(h) Extrapolation and interpolation criteria (if applicable).

The RCbS process commences at the same 'starting point', but aims to address these elements through flight simulation. A crucial step in the simulation requirements development as proposed herein is the identification and description of the flight simulation Influence, Predictability and Credibility levels. These levels differentiate how modelling and simulation are to be used, how good the predictive capability needs to be and how credible the predictions are, particularly outside the domain of validation (DoV—see text box) where judgements are made based on extrapolation.

The Four Domains in RCbS

Using M&S to describe and predict flight behaviour, four domains are considered (Fig. 4.2). In the case of a whole aircraft, the domain concept is intended to encompass both the region of the flight envelope and the range of aircraft configurations relevant to the ACR. In the case of a component, or feature of the flight model or flight simulator, domain is intended to encompass the range of relevant describing variables and states. The four domains are defined as follows:

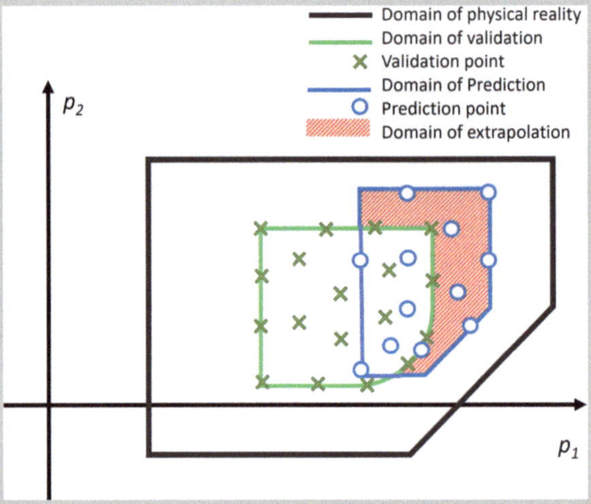

Fig. 4.2 Illustrating the domains concept in RCbS[1]

1. The domain of prediction (DoP); the domain within which it is the intention to predict the behaviour of the aircraft or component and to use these predictions to achieve certification at the defined Influence Levels for an ACR.
2. The domain of validation (DoV); the domain within which test data will be used to validate the flight model or simulator and their components/features. In Phase 2 of the RCbS process, validation implies a positive outcome has been achieved for the relevant metrics in the fidelity assessment. Within the DoV, interpolation is used to predict behaviour between validation points.

[1] FSM predictions are not always extrapolated from the DoV to the DoE; they may be computed without explicit reference to the DoV results. However, the modelled physics is conceptually extrapolated, if one considers the 'roots and branches' of fidelity lying in the DoV, growing out into the DoE; model updating parameters may be explicitly extrapolated of course.

3. The domain of physical reality (DoR) is the domain within which the laws of physics being used are considered to be adequately represented in the flight model and flight simulator. Since all models and simulations used in the RCbS process will include approximations to physical reality, this domain is strictly the region where the approximations are valid, reflecting the description 'adequately represented'. Of course, understanding the validity of approximations suggests a definitive knowledge of the DoR boundary. In practice this is hardly ever the case, so it is important to collect evidence that can show that the hypothesis underling the choices made to build the conceptual model are still valid. Sometimes this goal might be achievable by quantifying the error between the approximation and the results from a higher-order, more sophisticated, computational model.

   To maximise the confidence in the results of M&S, the DoV and DoE must both lie within the DoR. In practice, the RCbS process will often imply a lack of validation test data within the full DoP. So, a 4th domain is introduced.
4. The domain of extrapolation (DoE); the domain, outside the DoV, but inside the DoR, within which extrapolation of predictions are made to achieve certification at defined Influence Levels for an ACR. Activity in the DoE may include, e.g., high (safety) risk failure cases and controllability or stability assessments at extreme atmospheric or aircraft loading conditions. Another example would be the case where (physics-based) flight-model updating, proved to be successful in the DoV, is used in the DoE as part of the Phase 2 fidelity assessment.

4.2 Influence, Predictability and Credibility Levels

The ACRs, together with relevant rotorcraft engineering design data, form the basis for making decisions on the scope of the flight simulation to be developed for RCbS. The operational envelope, as part of engineering design inputs, defines the conditions under which the components of the FSM and features of the flight simulation may be exercised, and so the complexity of the physics to be modelled. Using this information, it is possible to define the required prediction domain of the simulation or FSM component, in terms of typical flight envelope parameters or FSM parameters. The prediction domain is one of four different domains relevant to the RCbS process, as described in the text box.

The description of Influence, Predictability and Credibility levels are used to convey meaning to the underlying consequences of the application of RCbS, in terms of safety and efficiency in the certification campaign. These descriptions form a foundation for the requirements capture/build process. The degree of Influence that the use of simulation will have on the certification decisions and the Predictability level anticipated for the flight model and simulator, and associated flight test system, will then impact the level of effort required throughout the entire RCbS process, as expressed in [3].

This Book takes a somewhat different perspective on Influence than the description in [3], where the focus is on how rigorously the NASA standard should be followed, with influence descriptors—negligible, minor, moderate, significant and controlling. In the present context, the levels of Influence on certification decisions similarly relate to the extent to which simulation is planned to be used in the certification associated with an ACR, but are described by the four options in Table 4.1.

In theory, a higher Influence Level will translate into more activity required in the development of model(s) and simulator(s), and in the validation, fidelity and Credibility assessments, providing increased evidence that the models are a correct and properly implemented mathematical translation of physical phenomena; that the simulation is a credible representation of real-word behaviour of the aircraft. However, there is potential for the Influence Level to be revised, upwards or downwards, following the fidelity assessment in Phase 2 or the Credibility assessment in Phase 3 of the RCbS process. It is, therefore, recommended that the full RCbS process is undertaken, where possible and appropriate, regardless of the initially selected Influence Level.

The plurals 'models' and 'simulators' are used above, highlighting that several variants may be used in the RCbS process. Some will be specifically tailored for application to an ACR, others suitable for more exploratory or de-risking tasks; some able to run in real-time, others coupled with computational fluid/structural codes. An

Table 4.1 Influence levels for use in certification by simulation

Influence levels		Description
1.	De-risking	The simulation is used to develop/familiarise with flight test procedures/techniques and to obtain an understanding of possible problems, hazards, or the need for additional data gathering etc.
2.	Critical Point Analysis (CPA)	The simulation is used to explore the flight envelope and aircraft configurations to be tested for a specific ACR, and to perform a down-selection of critical points to be tested in flight, yielding improvements in test efficiency and safety
3.	Partial Credit	The simulation is used to receive certification credit for a portion of the flight-envelope/aircraft-configuration matrix, or an aspect of an ACR. Supplementary flight tests will need to be performed to demonstrate full compliance with the ACR
4.	Full credit	This category is for cases where certification flight tests for a specific ACR are replaced by simulation

important aspect is then how these different 'versions' relate to one another that needs to be documented under 'configuration management' in the PMP (Chap. 3). In the following, the plurals are omitted but implied.

Within the DoP, the outputs of the FSM and FS are distributed throughout the DoV and the DoE. This distribution is described in terms of the Levels of Predictability, as illustrated in Table 4.2 for an example ACR. The Predictability levels are described in general terms as,

P1 Full interpolation: predictions performed only within the DoV, the (interpolation) errors for the quantities of interest can be estimated with high confidence,

P2 Extensive interpolation in the DoV and limited extrapolation in the DoE: note that all cases of acceptable extrapolation, as per the current CS-29 and CS-27 AMCs, are of predictability level P2.

P3 Interpolation in the DoV and extensive extrapolation in the DoE: including extrapolation beyond CS-29 and CS-27 AMC criteria, as well as for any design change that would otherwise require flight testing for supplemental compliance demonstration.

P4 Full extrapolation: all points used in simulated tests are outside the DoV and so no direct comparisons of the FSM or FS with flight test data are available.

Table 4.2 Typical layout for influence-predictability level matrix in the RCbS process

		Predictability Levels			
RCbS ACR	**Influence Levels**	Full Interpolation in DoV (P1)	Extensive interpolation in DoV Limited extrapolation in DoE (P2)	Interpolation in DoV Extensive extrapolation in DoE (P3)	Full extrapolation in DoE (P4)
ACR 29.143 (Controllability and Manoeuvrability) Control margins for low-speed manoeuvring in winds	De-risking (I1)				
	Critical Point Analysis (I2)				X
	Partial credit (I3)				
	Full credit (I4)			X	

The example in Table 4.2 shows that the FSM/FS is used to explore the DoE to identify 'critical' points (I2P4); e.g. those flight conditions where performance, control or stability margins are minimum, where failures lead to significant departures from trimmed flight or are difficult for the pilot to manage. Critical points may also refer to aircraft weight and balance configurations. From these results, a subset of conditions is selected for full credit, supported by results from DoV tests, hence I4P3.

Credibility and the Assessment of Confidence. Credibility analysis then considers the consequences to safety and performance from the reliance on simulation, for the assigned Influence and Predictability Levels. Credibility analysis informs an assessment of confidence, and vice-versa, and is particularly important for, but not exclusive to, test conditions in the DoE. So, for each ACR selected for RCbS, there needs to be such an assessment to determine the extent of flight test data required and the technical content, the complexity, in terms of features and components, of both FMS and FS. An approach to credibility assessment is introduced later in this Chapter. Prior to this, the following examples are used to illustrate the integrated nature of Influence and Predictability assessment, and how this feeds through into the detailed FSM/FS requirements.

(a) Appendix B of both CS-27 and CS-29 (Airworthiness Criteria for Helicopter Instrument Flight) quantifies the requirement for a helicopter's dynamic stability in terms of the damping as a function of oscillation period. The intention (Influence level) might be that partial credit is sought for this ACR by achieving credit for 50% of the airspeed-altitude envelope, for all weight and loading configurations. To achieve this RCbS influence level, it might be proposed that medium–high altitude tests at high speed are replaced by simulation, using the validation results and model-updating process successfully developed with data from low-medium altitude (DoV) testing. One question that must be asked is what FSM characteristics are considered necessary to ensure sufficient Credibility in this 'extrapolation' process. For example, previous experience may have indicated that with the aerofoil sections used on the certification aircraft, dynamic stall is to be expected at high Mach numbers, with consequent impact on blade torsional response and the damping of the aircraft pitch-heave oscillations. In another example, modelling correctly the loss of dynamic pressure at the vertical stabiliser, due to fuselage interference (blockage) effects, might be considered critical to capturing the reduction in weathercock stability and the impact on the frequency of the lateral-directional-oscillation. In could be argued that the model-update process embodying this effect, that was successful at low altitude, could be replicated at the high-altitude conditions.

(b) ACR 29.143 (Controllability and Manoeuvrability), requires that the "*wind velocities from zero to at least 31 km/h (17 knots), from all azimuths, must be established in which the rotorcraft can be operated without loss of control on or near the ground in any manoeuvre appropriate to the type.*" To avoid the safety risk in flight test, a combination of (off-line) CPA and piloted simulation might be proposed to achieve partial or even full credit for defining the flight envelope within which loss of controllability in such low-speed manoeuvres might occur (see Table 4.2). The FSM characteristics considered to be important in this application include the interaction of the main rotor wake with the ground, fuselage, empennage, and particularly the tail rotor, when hovering in winds from different directions. It is recognised that, at least for conventional rotorcraft configurations, to achieve sufficient fidelity at critical azimuths, high fidelity Computational Fluid Dynamics (CFD) or vortex-wake solutions are likely to be necessary. Converting the solutions into reduced-order models (e.g. data-maps) for real-time computations represents a significant challenge. An FS characteristic considered important might be 'realistic' fine-textured ground surfaces that provide the pilot with sufficient translational and attitude motion cues. It might also be proposed that the provision of vestibular motion cues is important, allowing the pilot to anticipate the visual motions. Likely characteristics required in the FTMS for the pre-certification flight testing to support validation are low airspeed pace-car trials with angles of incidence α and sideslip β sensing, and tail rotor flapping data.

(c) ACR 29.53(a) relates to Category A take-off requiring that a rotorcraft, following an engine failure, can return to and land safely in the (confined) take-off area, or continue the take-off and climb-out. De-risking might be sought by using a piloted simulation to evaluate the robustness of the defined take-off procedures, to explore variations in procedures and control strategies, and to determine the maximum take-off weight from energy-management and low-speed controllability perspectives for all foreseen helicopter configurations and within the applicable flight envelope. FSM characteristics considered important to the accurate prediction of power/torque limits, and transient one-engine-inoperative torque/rotorspeed response, include the heave/yaw responses to collective control inputs, the high-complexity rotor wake with strongly non-uniform radial inflow distributions, taking account of ground-effect, the thermodynamic characteristics of the engines and the functions within the engine and rotorspeed control systems. In the FS, establishing sufficiency for the fidelity of the visual and vestibular cueing will need attention, particularly in the final phase of the manoeuvre prior to touch down of the rejected take-off. Ensuring that the pilot-vehicle interface is representative of the aircraft is also important for achieving equivalent pilot control strategies. Obvious examples here include inceptor forces and displacements, instrument panel layout and detail and outside-world field-of-view. In the complex, steep descent phase or the approach to touchdown there may also be a concern that variations in control strategy could bring the aircraft close to Vortex Ring State (VRS); modelling the unsteady aerodynamics and nonlinear flight behaviour in VRS is notoriously difficult [4, 5], but is advised for this ACR in view of the safety implications.

(d) A fourth example is drawn from the multiple requirements relating to stability augmentation systems (SAS). CS 29, Appendix B, VII(a) requires that "*for any failure condition of the SAS which is not shown to be extremely improbable, the helicopter is safely controllable when the failure or malfunction occurs at any speed or altitude within the approved* (Instrument Flight Rules) *IFR operating limitations*." Initially, the FSM might be used to conduct a CPA of in-flight SAS failures throughout the flight envelope to establish conditions to be tested using an FS. The latter would then be used to achieve partial or even full credit, with a defined pilot reaction time, to demonstrate safe recovery and continued flight after the failure with representative cueing (aural, tactile, vestibular, etc.), and without the need for "*exceptional piloting skill or force*". Important features here are likely to include the ability to model correctly the effect that the SAS failure has on aircraft response, the failure cueing (e.g. accelerations), control forces and flight characteristics at unusual attitude excursions that might arise during the recovery phase. Depending on the failure, Hardware-in-the-Loop (HITL) testing with the actuators and/or Flight Control Computer might also be considered appropriate.

As part of the Influence-Predictability (I-P) Level assessment, an applicant needs to define, for each ACR for which RCbS is sought, how the activity is distributed throughout the DoP described above. This can be achieved in terms of the extent of interpolation (activity within the DoV) and extrapolation (activity within the DoE). Returning to Table 4.2, example (b) above is used to illustrate how the I-P Levels might be defined, by identifying the planned elements of Influence and Predictability.

However, the selection of I-P Levels requires additional quantifiers to establish the level of credibility expected from the results of the RCbS process. As noted above, Credibility relates to the confidence an applicant has that the results from M&S reflect the behaviour of the real aircraft. Several factors will impact this confidence and hence Credibility, for example;

(a) The M&S capabilities of the applicant, documented in reports and papers, international recognition of subject-matter-experts, including fidelity assessment and experience with model-updating methods.
(b) Extent of previous experience with the prediction of the specific behaviours related to an ACR, including on different types, and informed by understandings of the kind of physics required to match theory with test.
(c) The extent of extrapolation, i.e., how far outside the DoV the prediction conditions are. It could be argued that the confidence relating to a small extrapolation is no worse than that from a large interpolation within the DoV.
(d) Understanding of the way the flight-physics evolves from the outer boundary of the DoV to the boundary of the DoP. Such understandings can be derived from previous experience (see b) or from the results of M&S at various levels of complexity. Evolutions that feature strongly non-uniform, unsteady or non-linear effects should attract detailed scrutiny to establish credibility.
(e) Complementary with (d), how the extrapolated referent trends from within the DoV into the DoE evolve.
(f) The confidence in the underpinning flight model updating methods used within the DoV and extended into the DoE.
(g) A strong factor impacting credibility relates to the expectations of the analyst, based on experience and understanding of how the physics is represented in the FSM. Bringing expectations into the quantification is important but also carries

a risk. Prior experience may not be directly applicable to the new case and this needs to be reflected in the uncertainty analysis.

(h) Thorough understandings and quantifications of uncertainty are thus prerequisites of strong credibility.

Confidence is an elusive concept, but for RCbS it must be reinforced by quantitative analysis of the uncertainties in predictions, and test measurement data. Figure 4.3 illustrates the Confidence Ratio (CR) concept used in this Book to quantify the credibility assessment relating to the prediction of a 'margin'. M is the margin, or the generalised 'distance', between the performance requirement (e.g. control limit, touch-down velocity or the damping of an oscillation) and the FSM prediction, i.e. the performance assessment in Fig. 4.3. Credibility assessments are concerned with deriving, and ultimately ensuring, the sufficiency of the variety of margins related to an ACR.

In Fig. 4.3, U is the uncertainty in the prediction of the performance. A generalised CR can then be defined as,

$$CR = M/U \tag{4.1}$$

An intuitive result of this simple expression is that the smaller the margin to the performance limit, then the lower should be the uncertainty. Confidence then relates to the relative size of U and M, and from a safety perspective it seems appropriate to define a minimum acceptable CR for the credibility-safety trade-off. The uncertainty topic will be returned to with more detailed discussion in Chaps. 6 and 9. In Eq. (4.1), the components of U are the (FSM) output uncertainties, derived from propagation of the input uncertainties (see Fig. 9.2). Hence, both M and U must have the same units of course, relative or absolute e.g. % control margin or absolute kW of power margin. Later, in Chaps. 6 and 9, the various components of U are discussed in more detail; those 'stemming from' uncertainties in, e.g. the design data input parameters (u_{inp}), the test/experimental data, the so-called referent, used in the validation and fidelity assessment (u_r), or the numerical/analytical solution processes (u_{num}). To emphasise, it is the impact of such (input) uncertainties on the (output) uncertainty in

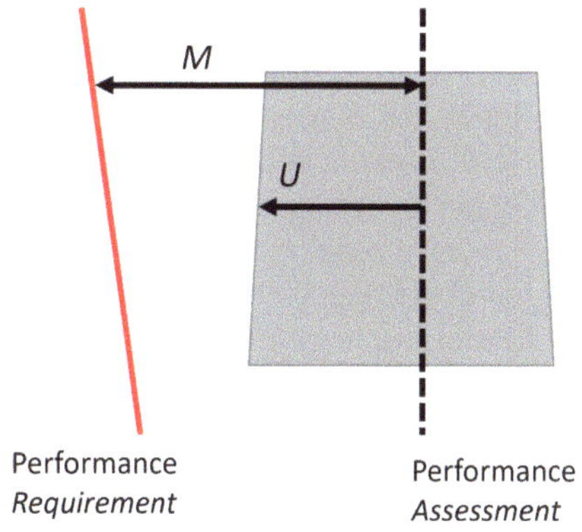

Fig. 4.3 Conceptualisation of the confidence ratio relating to a performance margin (M) and uncertainty (U)

the performance assessment/prediction that must be computed to form the combined U. Individual uncertainties will be designated with lower case u, and subscripts as indicated above. The way uncertainties at these sources are propagated through the model to impact the u's, and hence U, will be initially assessed in Phase 2 as part of the V&V activity (Chap. 6) and concluded in Phase 3 as part of the credibility assessment (see Chap. 9, e.g. Fig. 9.2).

In the above context, the *CR* concept only applies to situations where there is a quantified performance requirement for a given parameter of interest. In case of ACRs for which such a performance requirement is not readily specified, e.g. because it relies more on subjective pilot assessment, other criteria for assessing credibility will need to be agreed upon. In the case of subjective assessments, this might be related to opinion consensus between three or more test pilots, where it is required that the assessments agree as to the suitability of the FS being employed and the performance achieved. Of course, it must have been demonstrated in Phase 2 that the FS being used has sufficient (predictive and perceptual) fidelity for the ACR.

The RCbS fidelity assessment process in Phase 2 will rely on metrics (defined in Phase 1) for a range of parameters to quantify the match between simulation and test data in the DoV, and for deriving margins for the sufficiency of fidelity (the fidelity tolerances). Extending uncertainty analysis to the broader assessment of fidelity requires a different perspective, as discussed further as part of the Phase 2 FSM development in Chap. 6.

This broader assessment of uncertainty, in terms of fidelity requirements and predictions (acceptable errors between simulation and test), is strictly only applicable in the DoV, where test data exist. In the DoE, with no test data, fidelity judgments will be based on, (1) the applicant's confidence in assessments in the DoV as discussed above, (2) confidence in their ability to predict any 'new' flight behaviours unique to the ACRs being investigated in the DoE, and (3) the quantified uncertainty, u_p, in predictions in the DoE.

The minimum requirement for the performance metric assessment is for positive confidence, i.e. $CR > 1$. Note that $CR < 1$ implies uncertainty larger than the margin; a situation requiring further attention, should certification be sought for such cases.

For added assurance, values of *CR* in higher ranges could be used; e.g. as suggested in Table 4.3. Here, the uncertainty is reflected in the level of confidence an applicant will have in the FSM prediction of the margin; smaller uncertainty reflecting a higher confidence level. However, at this stage in the RCbS process, it is emphasised that the limits and ranges of these levels are purely illustrative and are not based on a rigorous assessment or theory. For example, an applicant might justifiably set the minimum acceptable confidence to be 1.05 rather than 1.0, accepting the consequent performance penalty. Further discussion on how M and U might be quantified can be found in Chap. 9.

The uncertainty U incorporates, amongst other terms, the extrapolated prediction error uncertainties, as derived from the fidelity assessments in the DoV. The model-updating process carried out in the fidelity assessment will address the sources and

Table 4.3 Possible confidence ratio (CR) ranges

1.0 < CR < 1.1	Minimum Acceptable confidence (MA)
1.1 ≤ CR < 1.25	Medium confidence (M)
1.25 ≤ CR < 1.4	High confidence (H)
1.4 ≤ CR	Very High confidence (VH)

extent of contributions to prediction errors and uncertainties. The trend in the evolution of errors within the DoV can be important for quantifying its extension into the DoE and hence the related uncertainties.

CR in the I-P matrix. Bringing the *CR* metric into the I-P matrix allows for requirements to be set on the minimum levels of confidence in the predictive capability of the M&S. An example is shown in Fig. 4.4 which is colour-coded with the levels of confidence suggested in Table 4.3. Once again, the example is purely illustrative, but conveys the idea that increased confidence is required in certain cases, e.g. when aiming for full credit in the DoE. In Phase 1, applicants should specify the expected/target *CR* for every I-P mix selected for an ACR.

What is acceptable, in terms of the distribution of *CRs* within the table for an ACR, will likely be a topic of negotiation between the applicant and certification authority. Following the above process will ensure that applicants address Credibility in Phase 1 of the RCbS process, setting the scene for the uncertainty analysis in Phase 2 and Credibility analysis in Phase 3. It is no exaggeration to state that developments in uncertainty and credibility analysis are likely to feature large as RCbS is increasingly exploited. In this context, the community-wide sharing of good-practice by early adopters is strongly encouraged.

At this stage in the RCbS process, this matrix would be used to inform the detailed description of the flight simulation Requirements Specification—the primary output of Phase 1. Combining the levels of Influence and Predictability with *CRs*, along with the data requirements from the FTMS, it is possible to quantify the scale of effort and resources required to proceed with the RCbS process. This will also inform any updates made at this stage to the RCbS Project Management Plan. It is considered important that applicants develop a good understanding of the technical requirements

RCbS ACR	Influence Levels	Predictability Levels with Confidence Ratios			
		P1	P2	P3	P4
	I1	(MA)	(MA)	(MA)	(MA)
	I2	(MA)	(MA)	(M)	(M)
	I3	(MA)	(M)	(H)	(H)
	I4	(M)	(M)	(H)	(VH)

Fig. 4.4 Influence-predictability level matrix with confidence ratios in the RCbS process

before firming up on the resource requirements; an obvious statement, but one that needs to be stressed.

The examples given in the paragraphs and Tables above relate to what we describe as specific requirements (i.e. related to an ACR). As touched on earlier, in Sect. 4.1, the general requirements, relating to flight behaviour throughout the flight envelope, also need to be captured, and how this might be achieved is outlined in the next Chapter.

4.3 Flight Simulation Requirements

Using, as input, information from the set of certification requirements, the relevant engineering design data, and outputs from the I-P 'levelling' process, described in Sect. 4.2, it will be possible to begin the flight simulation requirement-capture/build phase, to create the requirements specification. The objectives here are to establish:

1. The types of flight simulation to be employed, e.g. desktop 'off-line' simulation, pilot-in-the-loop simulation, or hardware-in-the-loop simulation.
2. The requirements in terms of characteristics that the FSM(s) and FS(s) must exhibit, and the associated predictive and perceptual fidelity they should satisfy.
3. The flight test data required to support validation, and consequent fidelity and credibility assessments.
4. Documentation detailing the proposed FSM/FS requirements as well as the associated rationale and/or justifications.

The requirements are composed of a detailed description of the characteristics and capabilities of the FSM(s) and the FS(s) and their components and features, and associated flight test measurements that are relevant to the RCbS process. It is recognised that applicant organisations are likely to have experience in developing requirements specifications for their products. The material in this Book is intended to aid applicants build on this experience, and to propose an approach that forms a solid foundation for the FSM/FS/FTMS development processes in Phase 2. The following properties are suggested for the framework and content of the requirements specification. To ensure no gaps or inconsistencies exist, all requirements should be,

a. Measurable—numerically quantifiable through a parameter or metrics
b. Unambiguous—clear, straightforward to interpret
c. Predictable—location in the prediction domain
d. Substantiated—drawn from known evidence
e. Traceable—with a specified, direct, association with ACRs or more general flight behaviour
f. Appropriate—able to discern quality in the intended application
g. Complete—covering all functions and operations included in the ACRs

Components and Features. Individual elements that can be distinguished by their function within the FSM are referred to as components in this Book. An FSM is then

Table 4.4 Example FSM component requirements

FSM, FS or FTMS component/feature	
FSM—Horizontal Tail (HT, Left)	
Function	• Generate aerodynamic loads on the left horizontal tail
Operation	• Active during all flight conditions • Updated every solution time-step
Data structures	• Location relative to aircraft reference • Tables of loads as function of local airflow magnitude and direction derived from wind tunnel and/or Computational Fluid Dynamics (CFD) tests • Incidence, sideslip ± 180° (interpolated from data at every 5°)
Inputs	• Aircraft motion (velocities) • Local (main/tail) rotor wake (velocities) • Local fuselage interference velocities • Atmospheric motion • Pitch control input (if moveable)
Outputs	• 3 forces, 3 moments
Interfaces	• Fuselage • HT-right
Constraints	• Rigid attachment except for pitch (if moveable)
Domains of prediction and validation	• Ranges of velocities, incidence and sideslip angles from wind tunnel tests relevant to the ACR
Fidelity metrics	• TBD
Driving requirement(s)	• Contributions to static and dynamic stability

created as a collection of linked components. The FS also has components but the term 'feature' is used to characterise the systems that provide the pilot experience (e.g. visual system). The usage of this term is well known from ICAO Doc 9625 [6] and EASA's Notice of Proposed Amendment (NPA) 2020-15 [7]. Components or features can be described in terms of the requirements they are serving, addressing functions, modes of operation, data structures, inputs and outputs, constraints and interfaces with other components. The prediction domain within which the component/feature should operate also needs to be specified, along with the relevant DoV. The textual example below is intended to be illustrative of functional and operational elements of an FSM component, but is, of course, not prescriptive.

> Each rotorblade is divided into N blade sections, spanning the radius, each with its own inertial, geometric and (optionally) elastic properties; each blade section is a component. The **function** of each blade section component is to generate a lift, drag and pitching moment in response to inputs from neighbouring rotorblade components and associated aerodynamic components, e.g. the atmospheric free-stream component, the finite-state inflow component and unsteady dynamic stall component. Typically, 2-dimensional aerodynamic data tables, derived from wind-tunnel or CFD experiments, provide the lift, drag and pitching moment, in coefficient form, as functions of incidence and Mach number. Three-dimensional effects could be included as a function of yawed flow angles. The DoP would be quantified by the functional variations, while the DoV would be defined by the variations of, e.g., incidence and

Table 4.5 Example FS requirements

FSM, FS or FTMS component/feature	
FS—Pilot's controls	
Function	• Pilot application and tactile feedback on cyclic, collective and pedal inputs
Operation	• Force-feel feedback system • Active during all flight conditions • Updated every solution time-step
Data structures	• Table look-up for forces as functions of control displacement and rates and time
Inputs	• Pilot limb movements • Trim control switches • Autopilot parallel actuators
Outputs	• Control rod/linkage motions to actuators • Control movements to autopilot
Interfaces	• Main actuators • Autopilot system and actuators • Pilot
Constraints	• Control stops • Actuation rate limits • Servo-transparency effects
Domains of prediction and validation	• Full range of control movements • Validation ground tests
Fidelity metrics	• TBD
Driving requirement(s)	• FS perceptual fidelity

Mach number, over which the aero tables have been drawn from test data. Through interfaces with neighbouring components, the blade section will typically **operate** at every time-step by transferring motions and forces to neighbouring (outboard and inboard) components, through a solution process that is defined, verified and validated. Such interfaces also need to be defined (e.g. joints, rigid, elastic), together with constraints, such as attachments to the rotor hub with flap-stops and pitch links.

Domains of Physical Reality and Validation. The degree of detail and level of complexity in the FSM aerodynamic modelling has proved important to achieving a level of fidelity appropriate for use in vehicle design, and this is expected to be even more the case for RCbS. Of relevance to fidelity are the range and limits of approximations used in the FSM that need to be defined, with supporting evidence, as part of the validation process. Such ranges/limits define the domain of physical reality of the components; e.g. in terms of compressibility or dynamic stall conditions. Accordingly, each component, or collection of components, will have their own domain of validation, encompassing conditions that might fall outside of the domain of prediction at aircraft level.

Examples of tabulating FSM Requirements. Tables 4.4, 4.5, 4.6 and 4.7 provide (non-exhaustive) examples for cataloguing the requirements in a form that captures

Table 4.6 Example FTMS requirements

FSM, FS or FTMS component/feature	
FTMS—Air data sensor system	
Function	• Provide measurements of (total) aircraft velocity, incidence and sideslip (V, α and β), pressure altitude, vertical speed etc
Operation	• Active during flight • Measured and recorded every (defined) measurement time-step • On-line computation of kinematic-consistent (with e.g. inertial/GPS system measurements) aircraft velocities
Data structures	• Location relative to aircraft reference • Calibration tables
Inputs	• Local air motion relative to sensors (typically pitot static tube measuring static and dynamic pressure, α and β vanes) • Computer algorithms processing raw measurements
Outputs	• Calibrated velocities, angles etc. at aircraft reference point
Interfaces	• Calibration algorithms
Constraints	• Physical limits of vanes, low dynamic pressure limits
Domain of prediction and validation	• Range of calibrated V, α and β • Wind tunnel tests • Low airspeed measurements require independent validation
Fidelity metrics	• Derived from kinematic consistency analysis • TBD
Driving requirement(s)	• Minimise referent measurement errors and uncertainties

the properties listed above. In the examples, the 'fidelity metrics' and 'driving requirements' are largely left blank (to be derived (TBD)), but are included for applicants to complete, when appropriate. This Book recommends that the documentation of such requirements be undertaken as a comprehensive process, provide a route to the foundations, to support diagnostic analysis, and ultimately decision-making concerning validation and extrapolation domains of the components and the whole aircraft. This approach to component-level requirements and fidelity can be extended to the whole FSM and FS, i.e. aircraft-level, for simulating conditions throughout the flight envelope, for application to de-risking for example. It is recognised that the results of component-level validation and fidelity assessment do not necessarily read across to the same at aircraft level; in some cases, the whole can be more deficient than any of the individual parts, e.g. as a consequence of integration and propagation effects. However, the 'weakest link' at component level, in terms of fidelity, might impact fidelity at aircraft-level to a disproportional amount and always deserves careful attention.

Table 4.7 Example FSM requirements

FSM, FS or FTMS component/feature	
Flight Simulation Model	
Function	• To provide trim, stability and response analysis and characteristics across the range of flight envelope parameters used in certification, i.e. the domains of validation and prediction
Operation	• Off-line, desk-top (including trim, linearisation analysis and time/frequency analysis) • Real-time operation in piloted flight simulator • Coupled with virtual pilot for off-line studies
Data structures	• Configured as a data-driven (possibly multi-body) dynamic system of integrated components each described by parameters, time-varying states and controls
Inputs	• Pilot (real or virtual) control inputs • External (atmospheric) disturbances • Outside world surfaces
Outputs	• Trim, stability and response characteristics of the whole aircraft • Individual component state variations as functions of time or frequency
Interfaces	• Atmospheric model • Pilot, real or virtual • Cockpit systems (for FS) • Outside world surfaces
Constraints	• Outside world surface constraints defined by undercarriage characteristics and external shape • Individual component constraints and limits • When FSM is operating, the approach to any constraint/limit should be flagged to the operator
Domain of prediction and validation	• Generally, all dynamic elements with natural frequencies up to, e.g. 30 rad/s and characteristic amplitudes within the component constraints/limits • DoV—range of trim, stability and response for which fidelity metrics are available from test data
Fidelity metrics	• Derived from aircraft-level dynamic response to control inputs • TBD
Driving requirement(s)	• Sufficient fidelity for the ACR

4.4 Summary: Phase 1

The activities contained within Phase 1, assembly of the RCbS Requirements Specification, are summarised below.

i. ACRs from the Certification Specifications are identified for RCbS,

ii. The relevant aircraft design data are collected together with related uncertainties (u_{inp}) that meet the maximum acceptable requirements defined in viii,
iii. the four domains within which the RCbS will be carried out are defined, (DoV, DoP, DoE, DoR),
iv. the Influence and Predictability Level matrices are defined for the selected ACRs,
v. preliminary description of expected complexity content for the FSM, FS and FTMS needed to achieve 'sufficient fidelity' for each of the selected ACRs,
vi. analysis and metrics for fidelity assessment, together with tolerances for sufficiency, are defined, in preparation for meetings with certification authorities,
vii. definition of test data requirements to characterise the DoV including programme for pre-certification flight trials and ground tests (Table 4.7),
viii. analysis and metrics for uncertainty quantification (e.g. maximum acceptable u_{inp}, u_{num}, u_r) and CR for credibility assessment are defined,
ix. Output; the (preliminary) RCbS Requirements Specification assembled based on the above, using a comprehensive descriptive framework.

References

1. Perfect P, White MD, Padfield GD, Gubbels AW (2013) Rotorcraft simulation fidelity: new methods for quantification and assessment. Aeronautical J 117(1189):235–282. https://doi.org/10.1017/S0001924000007983
2. Tischler MB, White MD et al (2021) AVT-296 Rotorcraft flight simulation model fidelity improvement and assessment STO-TR-AVT-296-UU. NATO STO
3. Anon (2016) Standard for models and simulations NASA-STD-7009A. NASA
4. Johnson W (2005) Model for vortex ring state influence on rotorcraft flight dynamics. NASA/TP–2005–213477
5. Brand A, Deir M, Kisor R, Wood T (2011) The nature of vortex ring state. J Am Helicopter Soc 56(2):22001. https://doi.org/10.4050/JAHS.56.022001
6. Anon (2012) ICAO 9625 manual of criteria for the qualification of flight simulator training devices. ICAO
7. Anon (2020) NPA 2020-15 Update of the flight simulation training device requirements, EASA

Chapter 5
Rotorcraft Flight Modelling and Simulation

5.1 Introduction

Following a requirements-based development approach ensures that the types of flight simulation utilised are appropriate to the selected ACRs, aircraft configurations and flight conditions of interest. Before progressing to the FSM/FS/FTMS building Phases, this Chapter discusses, in general terms, the range of flight modelling and simulation options available in the RCbS process; specifically, real-time pilot-in-the-loop, off-line desktop, hardware-in-the-loop, and flight with a virtual pilot, prior to the more detailed materials in Chaps. 6 and 7.

5.2 Simulation Type

In the RCbS process, it is likely that a family of FSMs will be used, ranging from moderate to very high levels of complexity. The decision as to which will be used for an application will be driven by the requirements and the selected ACRs. For example, if pilot-in-the-loop response characteristics and subjective pilot assessment of controllability and manoeuvrability are required, a real-time pilot-in-the-loop simulation, in a FS, will be required. Conversely, open-loop (stick-free or stick-fixed) handling qualities and performance analyses may typically be performed with a standalone FSM in an off-line desktop simulation environment. In certain cases, manoeuvre control by a virtual pilot (see text box) may be advantageous.

Predictive fidelity assessment with a virtual pilot forms an element of one of the Case Studies presented in Chap. 13. If pilot-in-the-loop simulation is required, it is necessary to provide not only the capability to simulate with a sufficient level of predicted fidelity the aircraft and the environment, but also all the elements of the flight simulator that contribute to the perceptual fidelity, e.g. visual and vestibular motion cues. But even here, the complexity of the FS depends on the application.

G.D. Padfield et al., *Rotorcraft Certification by Simulation and Analysis*, Springer Aerospace Technology, https://doi.org/10.1007/978-3-031-86398-1_5

Handling Qualities (HQs) and human-factors assessments require a cockpit and flight controls that are representative of the certification aircraft. In other cases, a generic engineering flight simulator may provide adequate realism to evaluate or demonstrate compliance and obtain partial or full credit in the conditions of interest. Decisions about the simulation type will be strongly informed by the requirements, which emphasises why the latter need to be sufficiently detailed, i.e. complete, substantiated, measurable etc.

The Virtual Pilot

This Book is not prescriptive about the form such a virtual pilot could take, but the intention is that the algorithms running such a computerised pilot need to be sufficiently realistic that meaningful conclusions can be drawn and relevant certification decisions can be made. One benefit of using a virtual pilot, or indeed any off-line analysis, is that the simulation does not need to run in real-time, so higher-complexity numerical (continuum-mechanics) models can be included, albeit that they also need to be verified and validated. A second benefit of using a virtual pilot is that a massive coverage of the whole flight envelope can be undertaken in batch-mode, isolating critical conditions for further investigation, for example through piloted simulation. The virtual pilot model may also be used to explore variations in defined procedures, e.g., evaluate the sensitivity of an emergency operating procedure described in the Rotorcraft Flight Manual (RFM) to realistic variations representative of the average pilot in operation.

It is recognised that, as the applications of RCbS evolve over the coming years, so too will virtual pilot models become more realistic and it is expected that their use in RCbS will expand considerably, for the reasons given above. With the current state-of-the-art in virtual pilot modelling, whenever the results are close to the certification limits, then a pilot-in-the-loop simulation is likely to be necessary for proof of compliance testing.

Relating to the selection of the simulation type, an element to consider during Phase 1, requirements capture and build, is whether or not the susceptibility of the aircraft to adverse rotorcraft-pilot-couplings (RPCs) or involuntary Pilot Assisted Oscillations (PAO) [1] (a more general form of the MIL-HDBK-1797A term 'pilot-induced oscillation' [2]) warrants specific off-line desktop analyses, or whether piloted simulation with appropriate kinaesthetic and vestibular cueing is required for the selected ACR. While the CSs and FARs only address this topic in a general way, the research to date suggests testing methods that can reveal an aircraft's susceptibilities to such PAOs [3, 4]. In addition, recent work on the susceptibility of tiltrotor aircraft to RPCs is discussed in [5], along with recommendations for testing techniques.

FSMs, particularly those used within a real-time FS environment, typically are not suitable for structural loads analyses, as may be required for proof of compliance with the relevant section of the CS (e.g. CS-29 sub-part C). Nevertheless, such simulations can provide inputs for more detailed off-line analyses using higher order (structural or aeroelastic) models, or may be used to evaluate high-level pass-fail criteria such as emergency touchdown sink rate and ground speed in lieu of a detailed landing gear loads prediction.

HITL simulation can be considered whenever there are vehicle subsystems, that the requirements suggest should be included, e.g., in the real-time simulation flight loop. Alternatively, it may be difficult to build reliable simulation models, due to complexities in the physics, and/or difficulties in collecting the data necessary, e.g. for commercial reasons. In such cases, the requirements on the inputs/outputs and interfaces between the simulator and the hardware components need to be clearly defined and conformity checks must be performed.

5.3 Strengths of Modelling and Simulation

Before progressing to examine the details within the three elements of Phase 2 in the RCbS process, the opportunity is taken to discuss some of the additional merits of using modelling and simulation in the certification process. Some of these may be well known and understood in the design and development departments, but might be less familiar to the testing community. The value to applicants of such discussion within this Book is that it can open-up new dimensions of awareness during the certification process, important for establishing the goals of the I-P matrix.

The real strength of the describe and predict capability of modelling and simulation is that it provides access to robust understandings of the connections between causes and effects; connections that are sometimes very difficult, or impossible, to make by examining the test data themselves (see text box).

Describe and Predict—mathematics in Action

Describe and predict are used to convey the fundamental purposes of modelling and simulation. For example, the trim analysis and solutions *describe* how the controls are used to achieve equilibrium flight conditions. In this sense, the word *describe* carries a general meaning. Trim analysis can also be used to *predict* the minimum-power flight speeds as a function of density altitude. In this sense, the word *predict* carries a specific meaning relevant to the application. Similar examples can be drawn from stability and response analysis.

One example of this strength is brought out through an examination of the changes in the forces and moments on components (or the whole aircraft) following a perturbation in a single state. A small perturbation in sideslip (or sway) velocity can reveal the changes in, say, aircraft roll and yaw moments. Increasing the magnitude of the perturbation can reveal the extent of any nonlinearity in these moments and deeper analysis can expose how different components, and their combinations, might be contributing to the nonlinearities. Ultimately, such analyses can then contribute to developing a full understanding of, for example, lateral-directional static and dynamic stability deficiencies. The latter may have been identified as a problem through piloted simulation, or even flight test, but the source of the problem could only be discovered through the kind of diagnostic analysis described above.

Another example of the strength of simulation comes from understanding the sources of problems relating to unexpected increases in main and tail rotor power, at both low and high speed. It is not unusual for poor power predictions to feature in the design and development phases, and often corrected by adjusting parameters in the rotor inflow model (low speed) or fuselage aerodynamic drag model (high speed). The importance of the correct physics in understanding poor predictions in the RCbS application requires a more clinical approach aimed at diagnosing the sources of mismatches between real flight and predictions. Increased effort will normally be required with higher-complexity FSMs and potentially increased pre-certification flight testing. The rewards of the increased efforts are likely to be found in both safety and efficiency during the certification process.

A third example addresses the whole gamut of de-risking, through the computation and management of 'large data' obtained from multiple, off-line, simulation runs, aiding the identification of cases for further analysis. The use of algorithms that can search throughout the flight/configuration envelope for boundaries of acceptable trim, stability or response characteristics or critical cases arising from failure modes analysis, can make such identification very efficient. The design of such algorithms is likely to feature large as modelling and simulation finds its place in the certification world.

The exploration, within the RCbS process, of 'what-if' type questions through off-line analysis and piloted simulations can also lead to discoveries that can impact the certification. Such exploration, without constraints, can be particularly valuable during the Phase 3 'Credibility' phase when extrapolation is under the microscope. We return to this in Chap. 9.

References

1. Pavel MD, Masarati P, Gennaretti M, Jump M, Zaichik L, Dang-Vu B, Lu L, Yilmaz D, Quaranta G, Ionita A, Serafini J (2015) Practices to identify and preclude adverse aircraft-and-rotorcraft-pilot couplings—a design perspective. Prog Aerosp Sci 76:55–89. https://doi.org/10.1016/j.paerosci.2015.06.007
2. Anon (1997) Flying qualities of piloted aircraft. MIL-HDBK 1797, US Department of Defence Handbook
3. Masarati P, Quaranta G, Lu L, Jump M (2014) A closed loop experiment of collective bounce aeroelastic Rotorcraft-Pilot Coupling. J Sound Vib 333(1):307–325. https://doi.org/10.1016/j.jsv.2013.09.020
4. Muscarello V, Quaranta G, Masarati P, Lu L, Jones M, Jump M (2016) Prediction and simulator verification of roll/lateral adverse aeroservoelastic rotorcraft-pilot couplings. J Guid Control Dyn 39(1):42–60. https://doi.org/10.2514/1.G001121

5. Padfield GD, Lu L (2022) The potential impact of adverse aircraft-pilot couplings on the safety of tilt-rotor operations. Aeronaut J 126(1304):1617–1647. https://doi.org/10.1017/aer.2022.20

Chapter 6
Flight Simulation Model Development (Phase 2a)

6.1 Introduction

Figure 6.1 illustrates the elements of the FSM development phase (Phase 2a), with major inputs from the Requirements Capture/Build Phase and the Engineering design data. Inputs from the parallel Phases 2b (FS) and 2c (FTMS) are also shown. The dashed lines indicate iteration pathways within Phase 2a and from outside, in both the parallel phases and the later credibility assessment in Phase 3.

6.2 Flight Simulation Model Build

Component-based adaptable-fidelity modelling. Put simply, an FSM used for certification compliance demonstration purposes should include the physics necessary to achieve sufficient fidelity[1] for the cases and conditions of interest, the ACRs. For a high level of confidence in the results, the FSM is applied within the DoV subset of the DoP. Beyond this, in the DoE, physics should guide the model content, and the levels of confidence in the results will depend on the credibility analysis introduced in Chap. 4 and expanded on in Chap. 9. The modelled physics shall describe the behaviour of the aircraft and predict the three essential aspects of flight, i.e. trim, stability and response (TSR). The FSM should, therefore, be *physics-based,* i.e., expressed in terms of, or derived from, the physical laws applied in the creation of the mathematical model and in the operation of the numerical simulation. The use of phenomenological sub-models for components is not considered prohibited, e.g. models not based on the 'laws' of physics but typically derived from test data. In some cases, full phenomenological models could be considered if P1 Predictability

[1] It is recognised that, in the Certification process, the term 'representative' is often used to mean sufficiently representative, i.e. two things are so alike that conclusions drawn from one are also applicable to the other.

G.D. Padfield et al., *Rotorcraft Certification by Simulation and Analysis*, Springer Aerospace Technology, https://doi.org/10.1007/978-3-031-86398-1_6

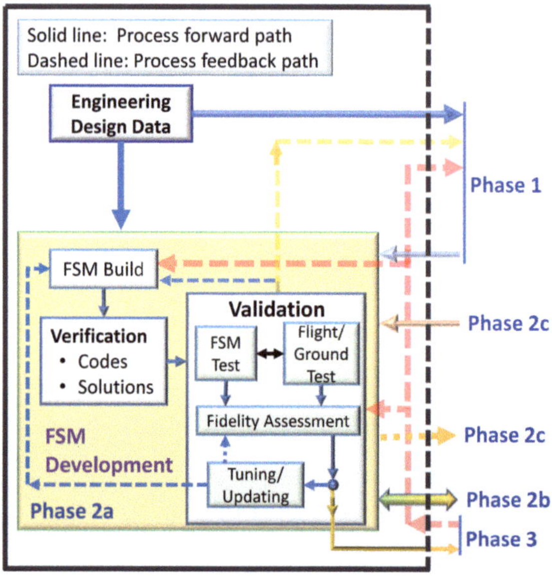

Fig. 6.1 The flight simulation model development; Phase 2a

level, i.e. interpolation only, is sought. However, for those cases, the identification of the associated DoR, and the assurance to not fall outside it, must be undertaken. The use of such models in critical applications can be considered as novel, at the time of writing, and should be highlighted to the certification authority. Preliminary guidelines are available for models derived from machine-learning in [1]. Early coordination with the certification authority is advised.

Appropriate (virtual) flight test monitoring parameters should be included in the FSM to ensure that it is not used beyond the limits of the DoR. Ultimately, the limits of validity of the FSM reflect the DoR boundaries; i.e. where the underlying model data and/or the mathematical approximations to the physics break down for modelling the various FSM components. These limits are expressed in terms of both global variables, such as those that define the flight envelope, and local variables, including the mathematical approximations to the physics being modelled for a component. Examples could be cited relating to the local unsteady aerodynamics of rotorblade stall or the dynamics of control actuators in saturation. Both have local and global consequences. The limits should be reflected in the DoV and DoR for the FSM, with the implication that the aircraft domain of validation should be within the ensemble of component limits of validity.

As stated in Chap. 4, it is not necessary that a single FSM be used throughout the RCbS process to perform all assessments for demonstrating compliance with the full Certification Specification. Multiple models, with different complexities and components, may be used, with the complexity driven by, and adapted to, the application. This adaptable-fidelity approach for certification is driven by the requirements on both the content of the modelling; especially but not exclusively for the aerodynamic forces, and the higher-order 'degrees of freedom' necessary to capture flight behaviour correctly. Examples are the importance of main rotor wake—tail rotor interactions for the assessment of low-speed controllability and the impact of dynamic stall on the blade torsion loads and vibration levels at high speed. This adaptable fidelity approach contrasts with the models typically used for training simulators which are required to have appropriate fidelity (from a training perspective, [2]) over the full flight envelope, but for which non-physical tuning may be acceptable.

Although other approaches may be conceived, a typical rotorcraft flight simulation model is composed of integrated components, or building blocks, assembled together, often following a Multi-Body Dynamic System (MBDS) logic. Figure 6.2 shows components that may be used in a typical helicopter simulation model. Similarly, Fig. 6.3 shows components used in a typical tiltrotor simulation model. FSM requirements for both types of rotorcraft will be similar although it is recognised that the latter are not certified according to CS-27/29 standards. A MBDS features multiple degrees of freedom (DoFs), represented typically in the FSM by component motion states and their velocities, or other states representing the evolution of the dynamic system (e.g. dynamic inflow states, pressures in actuators, turbine thermo-states). These would normally include the 6-DoFs of fuselage velocities, coupled with rotor flap and lag or gimbal motions, engine and drive train dynamics etc. In the current state of the art, finite-state rotor inflow models are typically used for real-time applications to capture local rotorblade section incidence and to estimate rotor wake interference on the fuselage and empennage.

Before turning the specific focus to fidelity, it is worth discussing a relatively new development relevant to the RCbS process. This Book advocates the application of physics-based modelling for the sake of the credibility of the simulation, particularly for extrapolated conditions. With the advent of Machine Learning (ML), it has become possible to derive rapid-execution, high-dimensionality surrogate, or phenomenological, models from test data and high-order physical modelling such as CFD. These data-driven methodologies such as Artificial Neural Networks, while not physics-based themselves, rely on training data that stem from testing or physical modelling. In principle, ML methods can be used for extrapolation beyond the training data set. The extent to which this can be done reliably and accurately for rotorcraft flight characteristics is still a subject of research. As such, this Book does not advocate such techniques for the first practices of RCbS, but early adopters are encouraged to investigate these avenues for exploitation in the future.

Required fidelity and sufficiency. To emphasise, the guidance in this Book advocates that the required fidelity, defined in the Requirements Specification for the relevant ACRs, is what is judged to be sufficient for the RCbS activity by the Applicant and the Authority, and evidenced through metrics across the DoV (recall Chap. 4). What exactly is 'sufficient' will depend on the application and it is one of the goals of this Book to provide guidance in this respect, in terms of DoV metrics and associated tolerances, for specific ACRs for which the RCbS process has been exercised. Nevertheless, expert judgement will be required to ensure that the fidelity is indeed sufficient for the particular configuration and application.

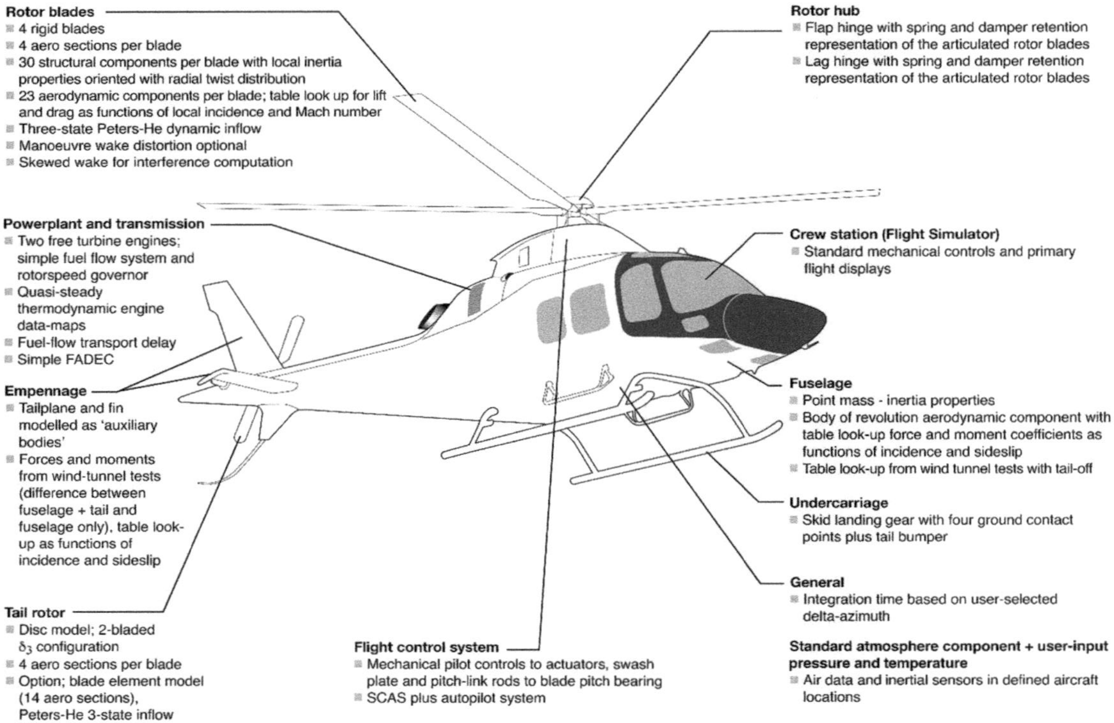

Fig. 6.2 A schematic example of a component-based helicopter FSM

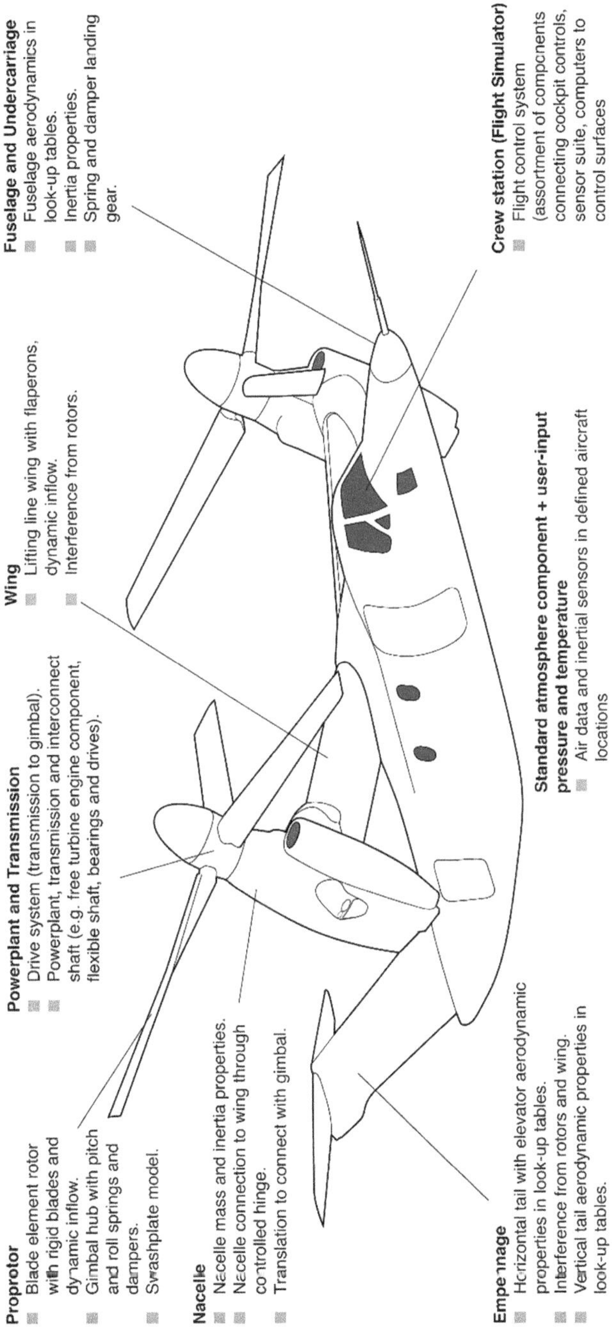

Fig. 6.3 A schematic example of a component-based tilt-rotor FSM

Interpreted in the 'frequency domain', an FSM operates over a wide range of frequencies and amplitudes and typically includes components whose 'natural' frequencies are higher than the range normally associated with classical flight mechanics and piloted flight control (0.1–10 rad/s), e.g. rotorblade aeroelastics, engine/rotorspeed/transmission dynamics, control actuators. The interaction of these components with the aircraft's flight mechanics and control can be important for achieving a 'sufficient' level of fidelity for certification purposes. In this context, a dynamic characterisation (e.g. eigen-analysis) can be carried out to identify the prediction domain in terms of the frequency ranges of component behaviour. Such characteristics are useful for establishing the excitation frequencies in the control-input designs used in the pre-certification flight tests to support FSM validation. Such results can also be used to compare with frequency response functions derived from test data. Metrics based, for example, on the 'allowable error envelopes' of the so-called 'maximum unnoticeable added dynamics' (MUADs) [3] can then, in principle, be used to quantify what can be considered sufficient fidelity. In this approach, the envelopes define the regions of response amplitude and phase within which the method predicts that a pilot will not be able to notice differences. Of course, the applicant must demonstrate that the associated envelopes proposed are applicable to the configuration and flight condition under consideration. Such caveats will apply to all fidelity metrics that require evidence-based justification to use in RCbS. In taking a frequency domain view such as described above, it is important to recognise that it may yet be required to include the higher frequency dynamics in the simulation if relevant for the ACR in question and even if it has no dominant role in the flight mechanics of the aircraft.

The degree of complexity in the aerodynamic modelling on all relevant components of an FSM has proved critical to achieving a fidelity sufficient for use in vehicle design, and this is expected to be the case for certification support. Of relevance to fidelity are the range and limits of aerodynamic approximations used in the FSM and these should be clearly defined, with supporting evidence, as part of the definition of the domain of physical reality and the related validation process.

An appropriate metric should be used to measure the fidelity sufficiency of any specific parameter (or combination of parameters), establishing a minimum accuracy though a 'tolerance' (indicated as '*tol*'). Only when the errors are below the tolerance can the fidelity sufficiency for that specific parameter be considered as achieved. The definition of this tolerance is important also for the experimental activities to be performed for validation purposes. Specifically, it will be important that any experimental uncertainty (e.g. due to measurement scatter or process noise) is kept well below this tolerance, otherwise it will never be possible to validate the models, demonstrating that the error is below the tolerance.

Trim, Stability and Response. The level of FSM fidelity relates to the three aspects of rotorcraft flight dynamics relevant to certification, as discussed in Chap. 4, namely (1) trim, (2) stability and (3) dynamic response (the TSR formulation). In general, the fidelity of the FSM for all three aspects should be 'sufficient' for the application, even if the application itself revolves mainly around a single aspect. That is, a model that accurately predicts trim for a control margin assessment, but poorly predicts short-term, open-loop, control response may be considered unsuitable for the former application. The FSM fidelity relative to flight test can be quantified for these three aspects, using appropriate metrics. Figure 6.4 illustrates a selection of metrics within the TSR framework.

Figure 6.4 was used in [4] to describe some of the parameters featuring in ADS-33 [5] to quantify a helicopter's handling qualities (HQ). Trim would be represented by the origin of the diagram, stability a function of frequency and response across the amplitude range, reflecting agility as a HQ parameter. Based on this description, examples of fidelity metrics might be:

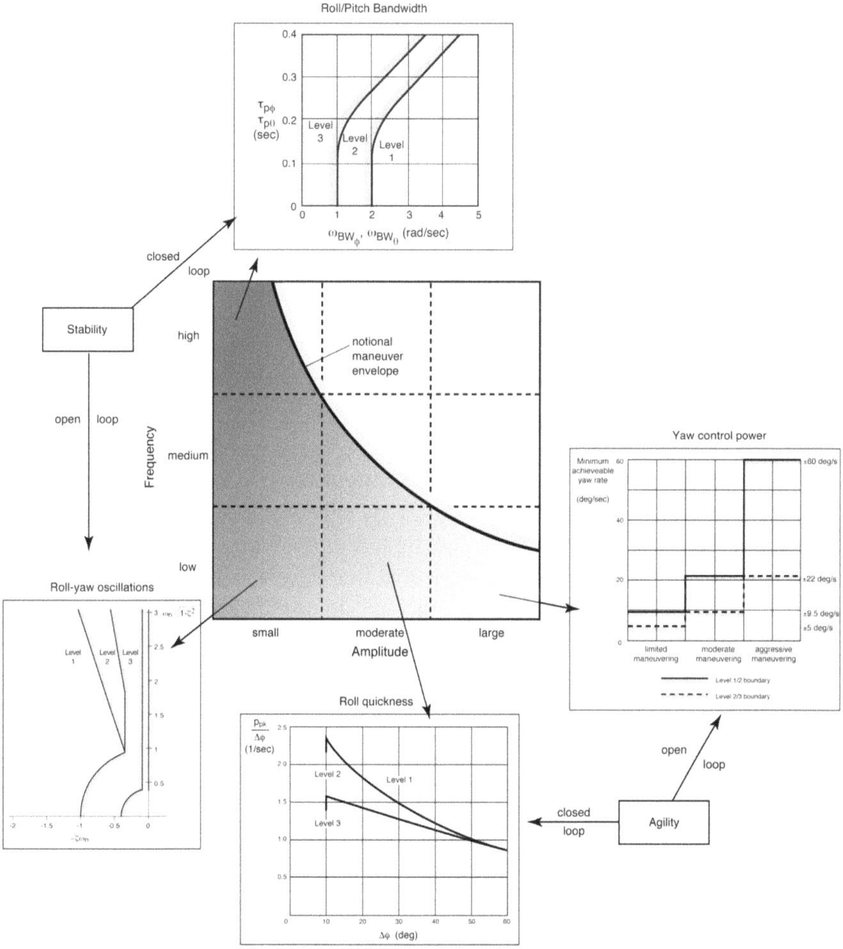

Fig. 6.4 Metrics for quantifying stability and response (manoeuvrability/agility) across the frequency/amplitude range of flight mechanics [4]

1. Trim; linking in current Certification Specifications with, e.g. the control margins throughout the flight envelopes and the static stability characteristics, the latter quantified in terms of control gradients for perturbations in lateral (sway) and forward (surge) speed. Trim computations are made with the non-linear FSM, to derive the control displacements required to ensure equilibrium flight, e.g. in terms of airspeed, sideslip, vertical velocity and turn rate.
2. Stability; linking in current Certification Specifications with the dynamic stability characteristics, quantified in terms of the damping and frequency. Techniques to establish dynamic stability typically involve the pilot/computer applying (open-loop, usually controls-fixed following input application) e.g. doublet-type, control inputs and allowing the free response to evolve sufficiently long that period and damping of the aircraft response can be computed. An FSM can also be linearised by perturbing each state and control in turn and creating a derivative-model, with the eigenvalues of the natural modes computed to provide stability characteristics. Alternatively, linear models can be constructed using system-identification methods [6] applied to both simulation and flight test data for comparison. The eigenvalues can be plotted on the frequency-damping chart for comparison with the open-loop response analysis using the non-linear FSM. Such comparisons can provide information on the impact on stability of any nonlinearities in the FSM. It is noted that the military standard, ADS-33E-PRF (ADS-33) [5] also includes requirements for 'closed-loop' stability in terms of attitude bandwidth and phase-delay as shown in Fig. 6.4.
3. Dynamic response (open- and closed-loop); linking, for example, with controllability and manoeuvrability requirements in the current Certification Specifications. A case might be the response of the rotorcraft to some control input or external disturbance (e.g. gust) computed as transient time-varying evolutions of the aircraft motions. Such cases are typically used in the validation process to establish the quality of the prediction of the short-term response. Another case could be the frequency response, computed in the form of transfer functions, e.g. pitch attitude response to longitudinal cyclic, generated through frequency sweeps on control inputs. Response analysis may also include the evolution of the system to triggering events, or even entire (emergency) procedures, e.g. failure of sub-systems of the rotorcraft. The link with the concepts of controllability and manoeuvrability also emphasises the potential value of pilot-in-the-loop simulation in the compliance demonstration process (Chap. 7).

The 'describe and predict' strengths of M&S can sometimes weaken because of difficulties in explaining the causes of flight behavioural characteristics through analysis with the coupled, multi-body, nonlinear, dynamic system. Linearisation can be used to provide insight in some cases, and it is considered worthwhile as part of this guidance to expand on this point.

The linearised derivative model has been used in aeronautical engineering since the very early days of aviation [7] to facilitate the understanding of complex aerodynamic phenomena. For example, trim gradients can be directly related to static stability derivatives. Furthermore, the nature of derivative variations with flight condition can sometimes reveal the source of, e.g. strong nonlinearities, instabilities or reductions in control margins. Aircraft response to small perturbations can be predicted using a derivative model for comparison with the nonlinear model to aid the investigation of larger amplitude response behaviour. Derivatives can also provide a microscopic view of FSM validation through comparisons with phenomenological models derived using system-identification techniques from flight test measurements. As discussed later in this Chapter, FSM fidelity can be improved through model-updating using physics-based 'delta' derivatives [6]. Rotorcraft differ from fixed-wing aircraft in many respects, including the number of DoFs required to describe behaviour. It is usually necessary to include rotor dynamics and rotor aerodynamic inflow, engine-drive train dynamics and, of course, the couplings between longitudinal and lateral-directional motions as standard. This multi-DoF model linearisation, therefore, results in a much larger system than the conventional 6-DoFs description. This formulation facilitates the investigation of important coupling effects that might be hidden in reduced-order forms. So, while the derivative concept is used above in the context of stability analysis, the application in FSM development can be far more extensive. Underpinning this point is that aircraft are generally designed so that cause and effect are linearly related. However, there are situations, commonly at the limits of the flight envelope or during large manoeuvres, where nonlinear behaviour prevails, e.g. actuation rate limiting, rotorblade dynamic stall and rotor-wake interactional aerodynamics in low-speed manoeuvres. Such phenomena can usually be revealed through comparison between results from linear and nonlinear models, a valuable exercise in the diagnosis of complex flight behaviour. Conditions under which linear approximations and reduced-order models are valid need to be understood to enhance confidence in the use of linearisation. Such understandings can also play a part in the derivation of confidence ratios for use in credibility assessment (see Chap. 9).

While the TSR approach provides a solid framework, the general question of how to relate the qualitative descriptions of dynamic response into quantifiable requirements for civil rotorcraft certification is a topic of ongoing research. Apart from stability, such metrics do not currently feature in the Certification Specifications. As encapsulated by Fig. 6.4, in the military handling qualities/performance standard, ADS-33 [5], the minimum acceptable dynamic response requirements are quantified in terms of parameters such as attitude quickness, control power and inter-axis couplings. Also, the comparison of dynamic response plays a role in the validation of the FSM for use in the certification of training simulators (e.g. FSTD(H) [2], see Chap. 7).

An important part of the development for an FSM relates to model-updating and tuning; i.e. the process of improving the fidelity of the FSM to ensure sufficiency for purpose, e.g. for the use in certification. Complementary fidelity assessments, using TSR analysis, can provide insight into the required 'physics-based' updates in this updating/tuning process.

Before progressing to examine the verification and validation processes relevant to RCbS, some details on specific components and interactions are discussed.

Flight control and automatic flight control system. The Flight Control System (FCS) interacts with the (real or virtual) pilot and may vary in complexity from a fully Automatic Flight Control System (AFCS) with upper-modes,[2] to a basic SAS or an unaugmented mechanical system. The pilot is not a component of the FSM (unless it is virtual), but is the centre-of-attention in the FS, as discussed in Chap. 7. And, as further elaborated on in Chap. 7, the data feeding through from the FSM to the FS is a key ingredient of the FS fidelity. The FCS is no different to any other component in the FSM, featuring functional sub-components, inputs and outputs and interfaces with, for example, the rotor systems, aircraft electrical and mechanical power systems and the inertial and air-data sensor systems, as well as the computerised autopilot and SAS. Likewise, simulated FCS sub-components will also have their own domains of prediction, validation, extrapolation and physical reality. The latter needs to take account of nonlinear dynamic characteristics e.g. displacement and rate limits in the electro-mechanical actuation systems, or backlash and stiction in control runs. The ways in which redundancies are achieved and managed is also an important function within the FCS. It is this aspect that becomes crucial when ACRs relating, for example, to AFCS failures are being considered for RCbS. The aircraft failure modes, effects and criticality analyses will have defined the kinds of failure that require recovery action, either by the system itself or the pilot. In this application, the FSM must therefore include these failure modes and the consequent system behaviour will be scrutinised in the validation process. For some failure cases, the absence of validation data, precluded for safety reasons, can place them firmly in the domain of extrapolation. As discussed in Chap. 4, credibility analysis for such cases, including deriving the related confidence ratios, becomes particularly important.

It should be noted that a significant amount of current certification flight testing is conducted with the AFCS engaged, when it is needed to show compliance with Instrument/Visual Flight Rules (I/VFR) handling qualities in CS-29, and in some cases in CS-27. For such cases, the AFCS failures also need investigation for their impact on the handling qualities. This does not reduce the importance of ensuring that the flight characteristics of the so-called bare-airframe FSM meet the required fidelity standards, regardless of how much the AFCS might suppress natural handling qualities deficiencies. Understanding the physics at work in the bare-airframe flight behaviour is considered vital in RCbS applications. This is particularly true for failure cases of course but also, more generally, to reinforce credibility of results in the DoE. The presence of an AFCS may, however, significantly reduce the impact of uncertainties related to the bare airframe, making the validation of the FSM more

[2] The term 'upper-modes' refers to all control modes designed to enable the pilot to fly, partially or totally, "hands off". So, upper-modes are effectively autopilot control functions, additional to the SAS functions. The term stability and control augmentation system (SCAS) is sometimes used to describe the combination.

straightforward. So, it is acknowledged that in some applications, e.g. AFCS upper-modes analysis and at P1 predictability level, the validation of the FSM with AFCS-on may suffice.

In many practical applications, the applicant might elect to use an exact software copy of the AFCS, or even perform hardware-in-the-loop (HITL) testing, to ensure the fidelity is as high as possible. However, this does not obviate the need for comprehensive V&V analysis for such hardware/software. In a similar vein, it is not uncommon to find that the autopilot or AFCS designs contain proprietary data, so that the exact details of the design are not available, even to the rotorcraft OEM, but only to the AFCS design organisation. This will typically require either black box integration or reverse engineering of the design, effectively a model creation, based on input–output data. The level of complexity can increase significantly if there are multiple functions drawing on the same controls, e.g. stability augmentation, response-type augmentation, load alleviation or flight envelope protection. How the FCS performs when multiple constraints are approached simultaneously should have been addressed in the design specification, produced by the rotorcraft manufacturer. However, such behaviour needs to be verified (does the behaviour meet the design specification?) and validated (is the design specification correct?). Again, the validation of the complete system can be inhibited for flight safety reasons, emphasising the importance of credibility assessment.

Engine, rotorspeed and transmission dynamics. A similar situation, through the potential unavailability of commercially restricted data and the need for reverse-engineering, can arise for the engine component. The main function of the engine is to deliver the required power and torque to the rotors in trims and manoeuvres. The engine also provides power to the mechanical/hydraulic and electrical systems and, how this distribution is managed, particularly when close to the power limits needs to be modelled correctly. The inputs to the engine model include fuel flow, with rates controlled by the rotorspeed governor, and air from the atmospheric component. The engine intake will 'shape' the airflow from the atmospheric model into the compressor stage of a turboshaft engine. These details may point to the need for including models of the thermodynamic processes within the engine, developing through the four thermos processes in the control volumes of the compressor, combustor, gas generator and power turbine. Options here include a fully unsteady dynamic combustion, or a quasi-steady look-up table for the thermodynamics of the compressor-combustion processes. These details will be very important in edge of the envelope performance analysis, but for stability and dynamic response prediction it is the rotorspeed response and torque reaction on the fuselage from both main and tail rotors that are paramount. This is especially true in the simulation of engine failure conditions where it is vital to predict accurately (or at least conservatively) the torque response and available power of the engine(s). The reverse engineering process often assumes a state-space model with variable parameters (e.g. functions of power, hence nonlinear) defining the time constants and gains relating rotorspeed to torque and torque to fuel flow. Elasticity in the drive shafts will impact the natural frequencies within this component and may need inclusion if these fall within the range defined as necessary in the requirements specification.

Environment modelling. Creating a standard atmosphere model is a straightforward task, with air density, ambient pressure and temperature being a function of altitude. The flight simulation validation envelope will be defined by the flight test conditions required and achieved within this atmospheric model, with simulated test points adjusted for the real-world test conditions. Complications occur when, during flight test, the air is moving relative to the Earth's surface with potential vertical and horizontal shears/gradients. Such 'winds' are usually unsteady, containing gusts and turbulence, whose characteristics vary with altitude and proximity to the terrain and objects. Humidity levels impact the density, noting that humid air is less dense than dry air, and the modelling is even more challenging when precipitation occurs. Granted that it is best practice (and at times a firm requirement) to perform flight testing in calm weather conditions, the atmospheric model that forms part of the FSM may need to feature the primary effects of these complications, if they are suspected to have significantly affected the data gathered during the pre-certification flight test campaign and a basis for validation of these effects is available from measurements. Random turbulence can, in principle, be modelled based on extracted air-data system measurements in trim, prior to test inputs being applied. More structured unsteadiness, for example non-idealised gusts, are more difficult to model and can appear as so-called 'process noise' on the signals from the sideslip and incidence vanes. Developing a method for addressing the impact of atmospheric unsteadiness within the RCbS may prove to be important. Moreover, whereas flight testing in closely controlled unsteady atmospheric conditions is not possible in practice, flight simulation, piloted or otherwise, offers the ability to explore the aircraft response to well-defined unsteady conditions in a repeatable manner. For example, the impact of turbulence on controllability and pilot workload is known to be potentially significant and piloted simulation provides the opportunity to quantify this impact directly (see Chap. 14 for impact on dynamic stability), rather than, e.g. by requiring a minimum level of residual control authority in trim. Yet another example in which unsteady environmental effects may be important is in the assessment of pilot workload in failure conditions.

It is noted that, although the atmospheric simulations may be external to the FSM, other elements that fall under the category of 'environment', include ground contact/friction and obstacle air wake models, Air Traffic Control (ATC), and other airspace users. The relevance of these will depend on the application scenario and, in case of the latter two elements, whether or not pilot workload is an important aspect in the evaluation being undertaken.

Coupling with the flight simulator (Phase 2b). As shown in Fig. 6.1, a primary 'artery' from the FSM (Phase 2a) to the FS development (Phase 2b) is a validated FSM. However, the FS will need a 'prototype' FSM prior to the formal release at the end of Phase 2a. This would be used to support the development of many of the FS features (if not already developed and verified in prior activities) for which the fully validated FSM is not critical. The characteristics of such a prototype need to be defined within the FS requirements to ensure that Phases 2a and 2b evolve efficiently. Inevitably, the use of piloted simulation in support of partial or full-credit Influence Level will need to await results of the FSM fidelity assessment. To emphasise, certification tests using the FSM and FS take place after the credibility assessments in Phase 3, but there are obvious efficiency benefits in Phases 2a and 2b being completed at the same time. Fidelity assessment will define the DoV within which the predictive metrics demonstrate sufficient FSM fidelity for application to the ACR. Consequent FS fidelity assessments will then focus on ensuring that the FS is also suitable for the ACR. These assessments will normally address non-FSM related features, such as visual and vestibular motion cueing, pilot inceptor control forces and cockpit ergonomics critical for flight-related ACRs. However, it may be that pilot feedback, as part of the perceived fidelity assessment, draws attention to some FSM characteristics related, for example, to trim or dynamic response. The FS verification and validation processes should ensure that the FSM is coupled into the FS such that its behaviour is essentially the same as the standalone version documented in the output from the FSM development Phase 2a. The pilot perceives the FSM outputs through the filters of visual and vestibular motion cueing systems in the FS, which themselves need to pass through V&V processes. It is not unusual for a pilot to 'blame' the flight model for a FS fidelity deficiency, even though the FSM fidelity assessment has been successfully 'passed'. There are multiple ways that FS features can contribute to perceived deficiencies that need to be thoroughly investigated before re-visiting the FSM fidelity assessment. This topic is returned to in Chap. 7.

6.3 Verification and Validation

Introduction. As described in previous Chapters, flight simulation attempts to replicate the relevant aircraft flight physics using computer models and, if applicable, simulator hardware. Within the FSM development process illustrated in Fig. 6.1, Phase 2a, sit the two 'checking' processes that ensure that the FSM build is verified (all model requirements are correctly met) and validated (the model requirements were correct, and the replications are successful). The V&V processes are so vital to the success of RCbS that some additional 'conceptual' background is considered useful in this Book as build-up to the discussion on V&V.

The creation of a simulation in the generic sense may be represented through a triangular process where on one vertex is positioned the real system of interest, on the second vertex there is the conceptual model, i.e. the collection of assumptions and abstractions applied to develop a physical model of the system of interest, and on the third vertex is placed the computational model. In turn, each of these boxes is composed of several sub boxes that detail the different activities, shown in Fig. 6.5.

The development process starts at the system of interest, in this case the rotorcraft, or a subsystem thereof. Through requirements analysis, considering the operational conditions under which the system needs to be analysed/simulated, the accuracy

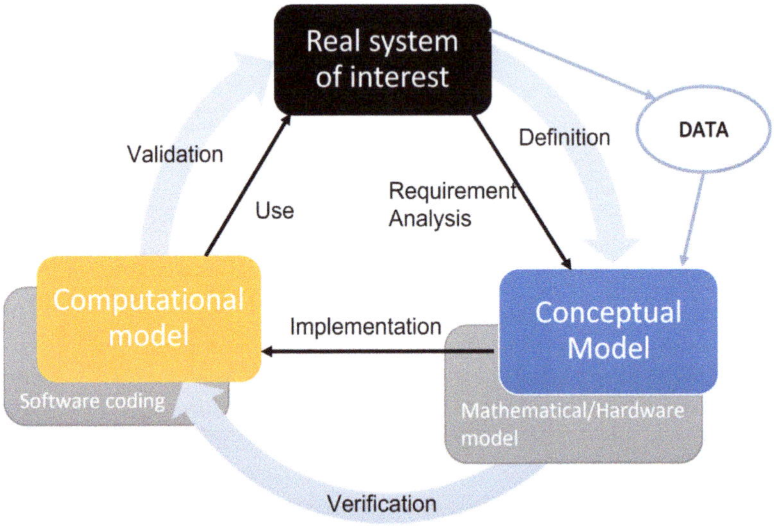

Fig. 6.5 Process to create a simulation model

required, etc., a conceptual model can be defined. This model can be transformed into a mathematical form, which is converted to a computational model with appropriate numerical discretisation, and finally implemented in the form of a computer code.

It is important to be able to trace back through all these steps. This, together with the knowledge (and related uncertainty) of the source of the data that are included in the model, will provide the information required to define the domain within which the model should function and operate correctly, i.e., the *domain of physical reality*, the DoR; and with sufficient fidelity within the DoV.

For physics-based models, the assumptions that define the conceptual model focus on what physical phenomena will be included and what will be ignored. As an example, consider the blades of a helicopter rotor as our real system of interest. Depending on the requirements, the blades can be represented as rigid bodies, as linear elastic beams, or more complex non-linear structures with sophisticated constitutive laws. Each of these conceptual models can then be represented by a variety of mathematical and computational model forms, each one with its own pros and cons that will make it more or less suitable depending on the requirements. At the same time, different sets of data will be required depending on the conceptual model choice. The models will have different DoRs, the limits of which depend on the nature of the loading applied to the blade. For example, the DoR for the structural aspects of an elastic blade element could be related to displacement and stress–strain amplitudes that satisfy linear constitutive equations. The DoR for the aerodynamic blade segment might be expressed in terms of the Mach number and angle of attack range included in the look-up data tables, combined with assumptions on local sweep angle and unsteady aerodynamic effects. Expression in terms of frequency and amplitude of dynamic response is another, general, approach to DoR description. However, since it will usually be difficult, if not impossible, to know exactly how far an approximation is from the DoR boundary, the best that might be achievable is a description based on comparison with a higher-order, more complex numerical model.

A similar process can be followed for the FS software and hardware. In this case, starting from the requirements, the necessary cues can be established for the pilot to acquire the correct awareness of the flight conditions, and up to what degree of realism they must be reproduced. This constitutes the bulk of the conceptual model for the FS. Then, it is necessary to define the software and hardware for the systems used to provide the cues to the pilot. Finally, as described in Chap. 7, the hardware/software systems are developed to translate the conceptual model into real cues, with associated DoRs for the FS features.

Once a simulation of reality has been built following the process outlined in this Chapter, the next crucial steps will be the V&V of the simulation. In the context of the above introductory paragraphs, *Verification* is the process of determining that a computational model accurately represents, within the required limits of accuracy, the underlying conceptual and mathematical models and its solution. The process can be divided into two steps: code verification and solution verification. *Validation* is then the process of determining the degree to which a model is an accurate representation of the real world from the perspective of the intended use. Usually, validation is performed by comparing the results obtained by the simulation with the results of experiments. However, in some cases, the reference data, or '*the referent*', as used in this Book, could be data, information, or knowledge gained by previous experiences, analogous systems, or even by other validated simulation models.

In basic scientific analysis, the predictive capability of a simulation model commonly deals with the ability of the underlying theory to be falsified by experimental observations. However, in engineering, the objective is to check to what extent the predictions meet the accuracy standards set in the requirements. So, the approach to validation is rather based on determining the acceptable level of disagreement between experiments and simulations. The level of disagreement is a measure of the fidelity of the simulation and validation revolves around defined fidelity metrics, and associated tolerances, that are used to quantify the degree of accuracy of the model.

Component-based building-block approach. It is advisable to undertake the V&V process, much like the FSM-build, in a hierarchical way starting from the simplest components up to the entire system that will be the object of the analysis, in this case represented by the aircraft. This can be described as a building-block approach, illustrated through a pyramidal structure as shown in Fig. 6.6. In this approach one moves from the lowest level, or tier, toward the top, increasing in complexity and the degree of coupling between different components. Validation at the lower levels is based on component-level experiments. In a general sense, rising through the tiers toward the top of the pyramid can obscure the coherence between causes (at low levels) and effects (at high levels), while the errors and uncertainties in measurements and relationships can increase, e.g. through propagation. The systematic step-by-step approach advocated in this Book should minimise the risk of this obscurity and ensure a higher control on the quality of the models both from a testing and analysis point of

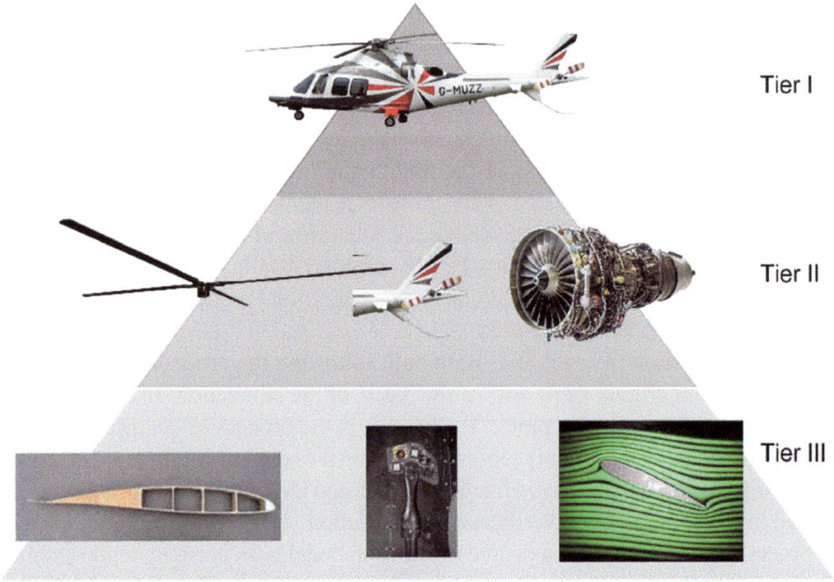

Fig. 6.6 Building block approach for an FSM

view. Such an approach also helps in isolating the cause of unexpected or erroneous results and should prevent modelling errors or deficiencies from being masked.

The guidance in this Book recognises the dual nature of top-down and bottom-up perspectives in V&V. In the validation process particularly, the comparisons of greatest interest lie at the highest, aircraft, level. However, mismatches here can normally only be understood, and ultimately resolved, at a deeper, lower-tier or component level. So, the two perspectives go hand in hand and a thorough grasp of the tier-connectivity is considered important in the RCbS process. Ultimately, the applicant shall make a credible case to the authority as to the lowest levels of model validation that must be considered for the application.

The requirements-based description of FSM development ensures that all interactions between components within and across associated tiers are fully defined. In the tier structure, this can be interpreted as couplings between the modelling in the adjacent tiers, as well as within a given tier. In principle, all the coupling should be two-way interactions. However, there are cases where the influence in one 'direction' is much weaker and might be neglected, or where the mathematical and/or computational formulations do not allow the implementation of a two-way coupling. In these cases, a one-way causal coupling may be the only available option, e.g. impact of undercarriage with surfaces. The consequences of the introduction of these one-way couplings must be thoroughly assessed, because they may limit the domain of the physical reality of the model and/or affect the component-level validation. Consider, for instance, the case where the model of a helicopter rotor is coupled with the modelling of the aerodynamic forces on the fuselage. The presence of the fuselage affects the rotor flow, but the rotor wake also affects the aerodynamic forces generated on the fuselage. Therefore, even though the isolated fuselage and rotor models may be validated for a large range of flight velocities and incidence angles, the range of conditions for which the prediction error of the coupled model is acceptable, as determined from validation at a higher tier, is likely to be narrower.

It is possible that not all the details of the subcomponents at the lower level of the pyramid are available to the applicant, e.g. because the aircraft is an assembly of subsystems developed and provided by third parties. This may be the case for instance for the engine, the landing gear, the avionic components, or the flight control systems. In these cases, if detailed modelling is required, it is advisable for the applicant to request a (traceably) validated simulation model from the subsystem providers, developed following requirements and conventions specified by the applicant and transparent in its formulation and implementation, so that it can be integrated into the FSM. If two-way interactions with the subsystem are essential, it is necessary to identify all inputs and outputs of the model required to perform the simulations. It is stressed that the requirements, in terms of V&V and documentation of the supplier subsystem model, are identical to those at the FSM Tier 1 level.

The application domains re-visited. In Chap. 4, the domains relevant to the RCbS process were introduced (Fig. 4.2). The domains feature large in the V&V and fidelity assessments in Phase 2a. While the definition of the DoP is relatively straightforward since it is derived directly from the requirements, the other domains require more detailed analysis to define. This is aided by Fig. 6.7 that illustrates, in 2-dimensional conceptual form, how the domains relate. Generalised flight envelope variables p_1 and p_2 are used on the axes, but it is noted that the domains are multi-dimensional and should be described as such by the applicants.

To define the DoR of the aircraft, it is useful to follow the building-block pyramid, as this relates the physics used on lower tiers to the coupled components at higher tiers. Typically, for each component at the base of the pyramid it should be straightforward to define the sets of inputs and outputs appropriate for the modelled physics. Then rising to the higher tiers, it is necessary to:

a. Identify the relationships between the inputs/outputs of the coupled model and the inputs/outputs of its components.
b. Identify the relevant input/output ranges of the coupled model that generate inputs/outputs of the component models that are within the respective domains of physical reality.
c. Ensure that the coupling physics in these ranges are correctly represented within the required accuracy. Alternatively, define the ranges over which the coupling effects could be neglected while meeting the requirements in terms of fidelity.

With the adoption of this approach, the coupled models should always have domains of physical reality that are equal to, or smaller than, the domains of the physical reality of the component models.

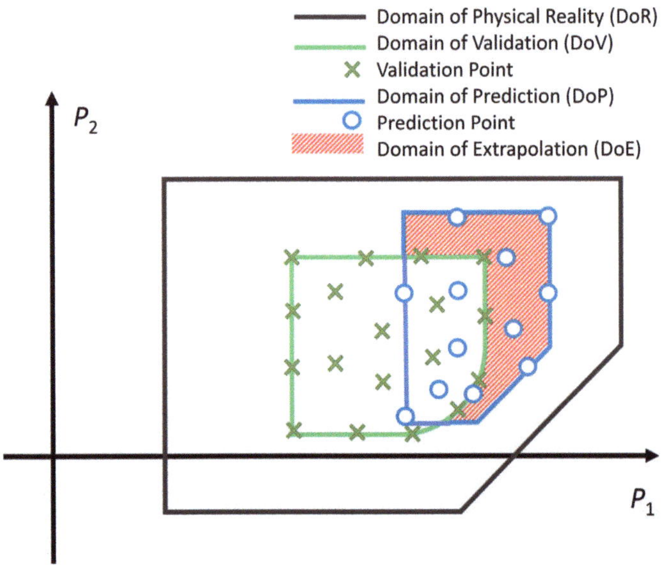

Fig. 6.7 Sketch of the application domains and their relationship

In some cases, elements of the building-block pyramid for an FSM may be composed of phenomenological (sometimes called surrogate) models, i.e., empirical models that are the result of the fitting of experimental data or predictions from higher-order numerical models. If so, care must be taken that these types of model are not used beyond their relevant DoV, i.e., the domain that contains all test data used to assess the fidelity and tune the coefficients of the model, e.g. [1]. This could be true for the top of the pyramid, if a model with only one level is developed. However, the logical consequence is that such a model will be used only for P1 predictability levels, i.e. interpolation only (Tables 4.1 and 4.2).

Although an initial/desired DoV will be defined in Phase 1, the 'final' DoV will be determined by the available test data and a result of the Phase 2 V&V process. It is, however, important to note that the DoV and DoP must always lie within the DoR. This is particularly important for the DoP, to avoid extrapolation beyond the limits where the model is expected to provide physically reliable results. Note that, in many cases, although a simulation might be operating in the global (aircraft-level) DoE, many of the components will be functioning in their local DoV (if such a Domain can be defined). However, the inverse may also be true. As such, it is important to trace how component models are operating relative to their local DoP and DoV boundaries. Component couplings are also relevant here, with dependencies enforced by physics-based couplings, e.g. rotor wake on fuselage loads, preserving the local DoPs of both.

Verification. The verification sub-phase assures that the implementation of the mathematical model through numerical algorithms is as intended. It is composed of two aspects, code verification and solution verification.

Code verification, establishing the correctness of the code itself is independent of the physical problem in the RCbS process. Code verification is concerned with ensuring that for a given set of inputs, the coded form of a modelled component generates the intended outputs. The first set of verification operations is described as numerical algorithm verification, usually performed by comparing the solutions computed by the simulation software with solutions generated by so-called *verification benchmarks*. These benchmarks can be: (a) manufactured solutions, (b) analytical solutions, or (c) numerical solutions appropriately generated. Applicants are referred to the literature, e.g. [8], for a thorough discussion on the different options. In general, cross-comparison against other verified codes is not considered the preferred approach, but in some cases, may be the only option. By comparing the computed solutions with the benchmarks, it must be shown that by reducing the discretisation size, the associated error tends to zero with an appropriate rate of convergence. By its nature, code verification must also follow the building block approach, starting from verification of components (e.g. a finite-element flexible beam element) and moving up to include component interactions (e.g. a rotating finite-element rotorblade with aerodynamic loading). Code verification for the complete FSM may, in practice, prove infeasible, reinforcing the importance of thorough verification at component level and associated couplings.

The second set of operations under code verification involves software quality assurance and pertains to verification that the code is reliable and produces repeatable results in a specified hardware and software environment. The bulk of code verification is often performed by the code provider. This is particularly the case when commercial software packages are used, e.g. NASTRAN, FLIGHTLAB [9], in which case, the applicant needs to collect evidence that verification has been undertaken by the code provider. This is considered a critical, although not always a straightforward, task, especially for large, commercial, multi-purpose codes. In addition, whenever the hardware or software is modified, it is important to verify that the 'new' code is still generating the results as intended. The latter requirement falls into the category of Configuration Management as addressed in Chap. 3. In some cases, it may be useful or necessary to ask the software providers to include the documentation to enable the applicant to repeat code verification tests. Finally, it is important that the impact on the FSM predictions of identified, or known, code defects are assessed as part of the solution verification process.

Solution verification is performed after code verification and is an activity with the objective of *estimating* the various numerical errors that stem from, for example, discretisation, iterative convergence and ensemble averaging of the FSM for a specific validation or prediction case. Through solution verification, as described in detail in Ref. [10] for CFD applications, it is the intention to assess the numerical uncertainty u_{num} of the model in the conditions being assessed. By its nature, u_{num} is epistemic, i.e. falling within an interval without an associated probability distribution, because the numerical algorithms considered here do not inherently carry any variability. Mathematically rigorous approaches exist to quantify u_{num} as an interval directly from the computer model, but these require greater effort, including gaining access to the source code of the model. A pragmatic approach is to consider the bounds of numerical uncertainty, obtained at a practical level of numerical discretisation, as defined equal to $\pm$ the magnitude of the error relative to the solution obtained at a higher, ideally asymptotically converged, level of discretisation. The numerical uncertainty is one of the ingredients for performing the fidelity assessment described in the following paragraph and, more particularly, the credibility assessment described later in Chap. 9.

It is also important to include all supplier and user-defined tools and scripts used to perform the pre- and post-processing activities in the verification process. In the case of time-constrained solutions (e.g. real-time application), there are also time-based convergence constraints, when the solution iteration may be halted before the established convergence criteria have been met. Requirements for how the solution process deals with such constraints need to be defined and the verification process should check that these requirements are satisfactorily met.

Validation and fidelity assessment. Validation involves the comparison of FSM results with a referent, which generally is the result of an experiment, commonly from flight testing in the RCbS process. The most common aircraft-level referent anticipated will be the prototype of the aircraft to be certified and data gathered during pre-certification flight tests. However, experimental referents may also be obtained from ground tests on components, from scaled models, noting that dealing with model scale test data presents additional challenges in terms of uncertainty quantification. Referents may also be derived from experiments conducted on an analogous system, e.g., flight test results from an aircraft variant that has a similar configuration. In some cases, experimental results may not be available so, other forms of referent can be considered, e.g. results obtained by higher-fidelity continuum mechanics models. However, this Book emphasises that comparison with such data is not strictly 'validation'. Such higher-fidelity modelling falls into the FSM toolset in RCbS and should receive the same level of scrutiny, in terms of V&V, as the core FSM.[3]

The goal of the validation process is to first compute the validation error and uncertainty, then to ensure the (updated) FSM predictions meet the tolerance requirements, defined in Phase 1, for the comparison of fidelity 'sufficiency' metrics with test data (see Chap. 4). Following on from this, the goal of credibility assessment is to collect enough compelling evidence that would convince, beyond a reasonable doubt, a group of peers that the predictions of an FSM are sufficiently correct. Credibility is returned to in Chap. 9. In performing validation, it must be remembered that experiments are also affected by errors and uncertainty and that these should also be assessed and quantified (see Chap. 8).

[3] This means, amongst other things, that the validation error and the uncertainty of the referent derived from the higher-fidelity model should be assessed and then used as the referent uncertainty u_r.

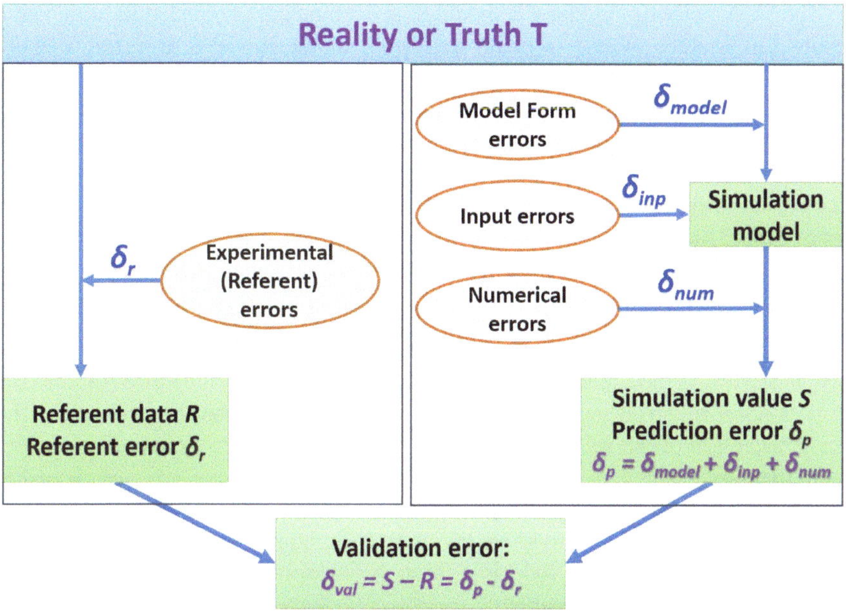

Fig. 6.8 Overview of the derivation of the validation (sometimes referred to as comparison) error

Validation error and uncertainty. Figure 6.8, derived from Refs. [10, 11], is used to support the description of key sources of error. So, we see that both the referent (typically experimental data) *R* and the simulation result derived from the FSM, *S*, have errors relative to the unknown 'truth'. The error in the referent value, δ_r, is the difference between the value of the referent and the truth *T*,

$$\delta_r = R - T \tag{6.1}$$

It is important to underline that the true value *T* is not known in practice, so in this case should be considered as an abstraction. Consequently, also the referent error δ_r is unknown and can only be characterised through the definition of an associated referent uncertainty u_r.

The error in the simulation result, termed the prediction error δ_p, and extending across the whole DoP, can be defined as the difference between *S* and the truth *T*,

$$\delta_p = S - T \tag{6.2}$$

The comparison between the simulation result and the referent is then called the validation error,

$$\delta_{val} = S - R = \delta_p - \delta_r \tag{6.3}$$

The V&V activities in Phase 2 should have eliminated, or reduced to 'acceptable levels', all sources of known error. Reduction of the model-form error is likely to be a major part of the validation process, as, for example, parameters in the rotor inflow dynamics, interference models or flap-torsion couplings are 'tuned' within their justifiable ranges. The remaining errors are not known and can only be described and quantified in terms of their uncertainties. The prediction errors that are the result of fidelity deficiencies (resulting in the outstanding model-form error) should then be contained within the uncertainties.[4] This issue is returned to in Chap. 9, when discussing credibility assessment in more detail. Referring to Fig. 6.8, the three elements of the prediction error, δ_p, for the simulation result S are (assumed independent, hence additive in their impact on the simulation prediction);

(a) the errors, in the simulation metric of interest, due to modelling assumptions, the model-form errors, δ_{model}, including those generated by the choices made in the conception of the model that are, by nature, related to epistemic uncertainties (EUns),
(b) the numerical errors, δ_{num}, in the simulation metric of interest, stemming from the methodology used to solve the underlying equations of the FSM, that are also epistemic, and are reducible through model refinement, even though this might not always be feasible,
(c) the errors, δ_{inp}, , in the simulation metric of interest, arising from the input parameters of the FSM. These errors may be related to epistemic or aleatoric uncertainty (AUns), or both at the same time.

Including the referent error, δ_r, the validation error δ_{val} in the DoV can be written in the form:

$$\delta_{val} = \delta_{model} + \delta_{num} + \delta_{inp} - \delta_r \tag{6.4}$$

Hence, the error due to modelling assumptions, i.e. the error an applicant needs to understand and contain within the defined tolerance in the DoV validation process, can be written as:

$$\delta_{model} = \delta_{val} - \left(\delta_{num} + \delta_{inp} - \delta_r\right) \tag{6.5}$$

[4] In this description of the RCbS process, valid is a 'definition' based on achieving a match between test and FSM within the sufficiency tolerance. The u's of the test data and FSM casts a shadow on the representativeness (in EASA terms) and need to be reduced so that the combined impact of mismatch and uncertainty still provides sufficient fidelity. For statistically quantified uncertainties, there needs to be an agreement on acceptable probabilities, as there will always be cases where a finite but small probability will push the fidelity outside sufficiency. The emphasis on ensuring acceptable low levels of the u's in Phase 1 will always be the best (and only) way to guard against ending up with more than small uncertainty.

To emphasise, a distinction is made between the validation error δ_{val} (and associated uncertainty u_{val}), defined with respect to the referent, and featuring only in the DoV, and the prediction error δ_p (and associated uncertainty u_p), defined with respect to the (unknown) truth (Fig. 6.8.).

The term in brackets on the right of Eq. 6.5 is composed of errors that are of unknown magnitude and sign. To progress from this conundrum, the errors are transformed into uncertainties. To begin with, and again assuming the errors are effectively independent (see Ref. [10]), which is not always the case of course, the validation uncertainty, u_{val}, can be defined as:

$$u_{val} = \sqrt{u_{inp}^2 + u_{num}^2 + u_r^2} \tag{6.6}$$

which, in the case of pure epistemic uncertainties, represents the best estimate of the combination of input, numeric and referent uncertainties. Statistical approaches, more sophisticated than the root square of the sum of squares, but beyond the scope of these guidelines, can be used to combine the different errors and obtain the standard deviation together with a more complete uncertainty structure of the error (see Refs. [11, 12]).

Although the 'heart of the matter', the simulation model-form error δ_{model}, cannot be uniquely quantified, it can be bounded to lie within the range:

$$\delta_{model} \in \delta_{val} \pm u_{val} \tag{6.7}$$

The measurement (referent), or experimental, uncertainty u_r is determined by the measurement set-up and reflects not only systematic and random errors in the data acquisition and instrument calibration, but also random errors or deterministic variability due to, e.g., atmospheric conditions, aircraft vibration or piloting technique, sometimes referred to as 'process noise'.

The numerical uncertainty u_{num} is obtained from the solution verification process, discussed earlier in 6.3 and in more detail in Chap. 9. Finally, the input uncertainty u_{inp} can be derived through uncertainty quantification (UQ) methods, also discussed in Chap. 9 as part of the Credibility assessment. It is important to emphasise that input uncertainty can, and should, be minimised through close collaboration between the design, manufacturing and simulation engineers in Phase 1 of the RCbS process. So, although the simulation/prediction and referent errors remain as unknowns, their uncertainties can, in theory, be quantified and used to bound δ_{model}. Setting requirements on minimum acceptable uncertainties is clearly a companion activity to setting tolerance requirements on the fidelity metrics in Phase 1.

To estimate an interval within which δ_{model} falls with a given degree of confidence (Eq. 6.7), an assumption about the probability distribution of this error can be made, if it can be justified. Defining U_e as the combined or 'effective' uncertainty used, it is possible to choose a confidence interval so that,

$$U_e = k u_{val} \tag{6.8}$$

where k is called the coverage factor. For example, considering purely AUns, if a Gaussian distribution is considered, taking $k = 2$, a level of confidence of 95% is obtained, that the true value lies within 2 standard deviations of the mean.

Following the approach of the ASME guide (see Sect. 6 of Ref. [10]), different scenarios can be envisaged, based on different relative values of the quantified δ_{val} and u_{val}. In any case, it should be verified, at least implicitly, that δ_{model} is within the tolerance, *tol,* to ensure sufficiency of the fidelity; which means that,

$$|\delta_{model}| < (|\delta_{val}| + u_{val}) < tol. \tag{6.9}$$

This should give a sufficiently robust indication that the modelling hypotheses is correct and that the elements neglected, or the approximation errors, have an acceptably low influence on the simulation prediction.

Consequently, to be confident that the model-form error is limited within the tolerance, it may be necessary to reduce the validation error, the uncertainty or both. In the favourable cases, where uncertainties and errors are within tolerance, to reduce δ_{model} two scenarios, and consequently two different strategies, are possible. The examples given below are for illustrative purposes and do not capture all possible scenarios.

Scenario 1. If the validation error δ_{val} is significantly larger than the validation uncertainty u_{val}, then it can be assumed to be dominated by the error due to modelling assumptions, δ_{model} (see Eqs. 6.4 and 6.7). Depending on the magnitude of δ_{val}, , relative to the fidelity sufficiency tolerance, this condition may be a symptom that something is missing in the model and it may even be necessary to return to the conceptual modelling activity (Fig. 6.5) and re-consider including physical phenomena that have been initially neglected. For example, it may be that rotorblade flexibility is initially considered to have a negligible influence on flight mechanics, and so is not included in the FSM. However, there are cases where the correct prediction of flight behaviour may require reconsidering this choice, e.g. flap-torsion couplings. In other cases, δ_{val} may be comparable to, or within the fidelity tolerance, in which case the model-form error is considered under control in terms of fidelity sufficiency.

For the case, where $\delta_{val} > u_{val}$, and $|\delta_{model}|$ is within the sufficiency tolerance (Fig. 6.9), the effective uncertainty, U_e, that features in the confidence ratio *CR* (Eq. 4.1), can be written as $U_e \approx |\delta_{model}| = |\delta_{val}| + u_{val}$. In Fig. 6.9. u_p is the uncertainty in the prediction error, δ_p shown in Fig. 6.8.

If extrapolation in the DoE is required, the extrapolation of U_e will require the implicit extrapolation of the model-form error. Predictions in the DoE requiring 'new' physics, or at least the exercise of physics present but benign in the DoV (e.g. dynamic stall effects), will then introduce further aspects to the model-form error that need to be considered.

Scenario 2. Alternatively, when $|\delta_{val}| \ll u_{val}$, i.e., $|\delta_{model} \pm u_{val}| \approx u_{val}$, no FSM improvements can realistically be formulated because the model-form error is within the uncertainty band imposed by u_{val}. . So, if the model-form error is not within the sufficiency tolerance, the focus should be first on reducing the uncertainty. However,

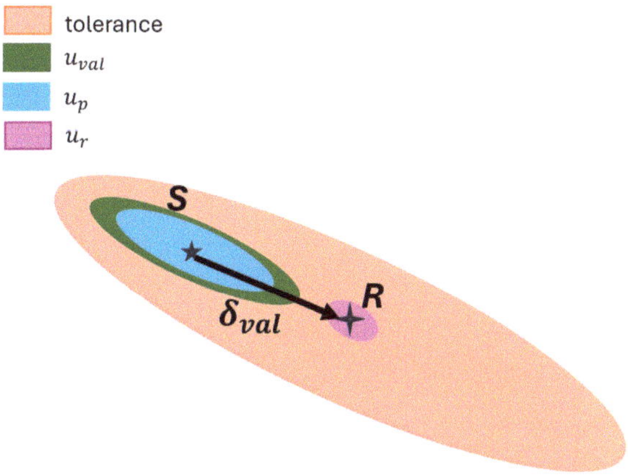

Fig. 6.9 Validation error and uncertainty. Sketch of the case where $tol > |\delta_{\mathrm{val}}| > \mathrm{u}_{\mathrm{val}}$. Note that the validation/prediction uncertainty range is centred on the simulation value, *S*, while the tolerance and referent uncertainty are centred on the referent value, *R*

if the (implicit) model-form error is within the tolerance band, i.e. $|\delta_{model}| < tol < 2\ u_{val}$, as shown in Fig. 6.10, the situation allows for a simpler quantification (and possible extrapolation) of the uncertainty $U \approx u_{val}$. This is based on the hypothesis that the relevance of the error in the DoV can be extended to the DoE, maintaining the increasing *CR* as a 'safeguard' for limited deviations from this hypothesis. However, any evidence to the contrary should lead the applicant back to Scenario 1. Since the model-form error is small, the only element of the effective uncertainty U_e is then composed of the prediction uncertainty that can be easily extrapolated. This is the ideal situation because the uncertainty U_e contains the effects of the 'known unknowns', while the potential effects of the 'unknown unknowns', related mainly to the model-form error, are limited. In this scenario, the sufficiency tolerance represents a target set for the validation uncertainty, and so a target for the combination of the different sources of uncertainty (set in Phase 1) that can drive the accuracy required when developing models and validation experiments.

Returning to the general principles of validation, although the level of detail in the fidelity and uncertainty analysis may vary depending on the ultimate FSM application, it is advised to follow the same hierarchical building-block approach as employed during model development for the validation process. Starting from a robust validation of systems at the base tier, typically characterised by a lower number of relevant parameters, the assessment at higher levels can concentrate on the interactional effects between the subsystems. FSM validation is thus recommended to be performed using this hierarchical approach, focussing on component-level tests and analysis before validation of whole aircraft behaviour against flight test data is attempted. The impact of fidelity deficiencies and uncertainties at component level on simulation outputs and the metrics of primary concern, often defined at the highest level, can be investigated through propagation analysis, an aspect of UQ that is returned to in Chap. 9.

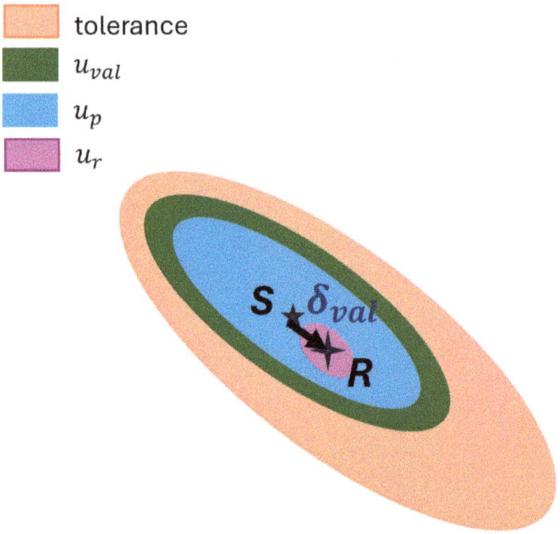

Fig. 6.10 Validation error and uncertainty. Sketch of the case where $|\delta_{model}| < tol < 2\, u_{val}$. Note again that the validation uncertainty range is centred on the simulation value *S* while the tolerance is centred on the referent value *R* so that it can be easily combined with the validation error δ_{val}.

Similarly, the validation of complex manoeuvres should be preceded by tests and analyses in relevant steady-state conditions and simpler manoeuvres. This type of bottom-up approach enables modelling deficiencies to be incrementally identified and, if possible, remedied. Here, support from diagnostic analysis of data/information flow through the FSM can be helpful.

Validation in parallel with development. An effective validation process will proceed in parallel with the development of the FSM itself and, also, the development of the experimental processes and systems, particularly the FTMS (Chap. 8). The availability of a prototype model before conducting validation experiments will enable the 'design of experiments' to progress efficiently, e.g. by selecting the relevant points to be tested and the quantities to be measured. This can also provide opportunities to solve problems related to a lack of information on the experimental conditions, or the quality of the different measures, etc., that are typically encountered when models are developed only after the completion of the test phases. The experiments dedicated to validating the FSM must be planned specifically for this purpose, identifying all quantities that need to be measured accurately to validate the FSM, which in some cases may be only loosely connected with the evaluation of the overall performance of the helicopter. These points are emphasised to alert the applicant to the importance of having the validation data-gathering an integral part of the FSM development.

Inherent test variability should always be considered and addressed to avoid the risk of relying on validation based on outlier data points, i.e. data points that fall significantly outside of the typical variability observed during a test or series of tests. Such cases can lead to large values of u_r and the consequent impact on credibility. Additionally, in these cases, the availability of a model before the test allows the applicant to build up knowledge on expected reference outcomes that can help in identifying possible outliers that require further investigation.

Validation flight testing. To relate to the Certification Specifications, flight tests for FSM validation should consider both steady-state (trim) and transient (stability and response) phases where relevant. Depending on the application, frequency-domain validation with associated system identification flight testing may be appropriate when the intention is to establish the veracity of the FSM across a frequency range. The bounds and tolerances for the so-called MUADs [3, 6] provide an example metric in this case. Alternatively, transient responses to control inputs can provide insight into short-term damping and control sensitivity comparisons, as well as cross couplings. The TSR framework for fidelity assessment was introduced in Chap. 4 and expanded on earlier in this Chapter.

When flight test data in the exact conditions of interest are not available or impractical to obtain, validation using interpolation is appropriate in the DoV. In the DoE, validation is strictly not possible of course, but a plan for the credibility assessment and related uncertainty analysis will need to have been laid out in Phase 1. The FSM will still be governed by the laws of physics in the DoE and the credibility assessment should determine how the flight behaviour might 'stretch' the modelling assumptions, possibly to their limits. Such analysis will be key to validation and fidelity assessment within the DoE.

Fidelity metrics. While an initial comparison between the referent and the FSM can be qualitative, to assess the correctness of the conceptual approach chosen, a quantitative analysis is required to determine the sufficiency of fidelity for the models used in RCbS. Building on the preparatory Phase 1 and Phase 2 TSR activities, it is possible to distinguish between different scenarios when defining metrics to quantify fidelity:

a. Cases where the objective is the evaluation of the trim value of a quantity of interest, e.g. a control margin. Here, the fidelity can be measured through the percentage errors between the referent and the result of the simulation. Ideally, the allowable error is referred to a meaningful performance characteristic related to the ACR, e.g., the minimum control margin for manoeuvrability or tolerance to atmospheric disturbances.
b. Cases where the objective is to compare the evaluation of a response over time, e.g. in dynamic stability assessment where period and damping are computed. Percentage errors across the time response, or metrics based on peak errors can be considered. The choice of metric should be influenced by the nature of the ACR being addressed.
c. Cases where the objective is the comparison of frequency response functions, e.g. to characterise and validate the frequency content of the various FSM components. The cited MUAD metric [3], suitably adapted to the application, has been established to connect with a pilot's perception of modelling errors across the frequency range of interest in pilot simulations.
d. Cases where the aircraft behaviour in an ACR involves large excursions from a trim condition, e.g. following failures or when pilots need to exercise the

full manoeuvrability of the aircraft, e.g. for obstacle avoidance. The moderate-large amplitude response quickness metric from ADS-33 is an option for fidelity assessment in such cases [5, 13].

In cases where there are multiple quantities of interest in the fidelity assessment, it may be appropriate to define a weighted sum of all quantities with the weights reflecting the relative importance of the different quantities [6].

As with all fidelity metrics, an applicant should provide evidence to support confidence in their veracity.

As illustrated in Fig. 6.1, validation and fidelity assessment should be considered an iterative process that starts with the fidelity requirements and associated flight test points (both defined in Phase 1) and advances to a tuned/updated model that meets the sufficiency requirements. The model updating process is also likely to be iterative, as fidelity at aircraft-level is derived from component-level assessment and developments.

It is emphasised here that the sufficiency criteria in terms of FSM fidelity in the DoV should be defined in Phase 1, based on the Influence/Predictability for the ACR and agreed with the certification authority. In principle, the 'acceptable' level of FSM mismatch-tolerances require a degree of engineering judgment based on the experience from other applications. For example, error tolerances from the Certification Specifications for flight simulation training devices may be used as a guideline where appropriate. More generally, a systematic approach is recommended to connect the physics-based nature of the FSM with the ACR.

Investigating FSM error. One of the challenges in the validation and fidelity assessment of an FSM lies in the systematic exploration of the model-form error δ_{model} and related (epistemic) uncertainties (EUns) in the predictions. That is, those errors and uncertainties relating to approximations in the mathematical expressions used to model the underlying physics. The following deals with the qualitative exploration of the model-form error, whereas quantification is discussed later in Chap. 9.

The engineer knows that their model is not perfect. S/he may have some knowledge of the imperfections and so should be able to estimate the boundary of the DoR, to try to ensure that predictions do not cross this. However, in several respects, the engineer will be rightly uncertain about the location of this multi-dimensional boundary. Also, in the DoR, and even within the DoV, s/he will likely have different degrees of uncertainty about the accuracy, the predictive ability, of their model and the precision of the referent data. Such uncertainties will be well informed by experience and, although experience is a good teacher, it can also deceive. As with most aspects of RCbS guidance, there is no substitute for a systematic process to assessing these kinds of uncertainties. Herein are suggestions for the elements of such a process.

Every input–output (mathematical) relationship in a FSM approximates reality. The mathematical approximations are reflected in assumptions about how the physical processes work. Here, for the benefit of clarity, we distinguish between (mathematical) approximations and (physical) assumptions. Understanding the physical assumptions and how they are represented by the mathematical approximations is part of the foundation for exploring the modelling errors in an FSM. This understanding is considered critical to the successful application of RCbS.

A significant benefit accruing from the use of an FSM, for both design and certification, is the ability to explore, identify and quantify the physical sources of the contributions to input–output relationships. Predictive capability has long been underpinned by understandings of the relationship between a cause and effect (the physical manifestations of input and output). In the limit, such cause-effect analyses can be used to quantify the impact of all modelling assumptions, and their approximant parameters, on mismatches between test and FSM predictions, i.e. the fidelity of the FSM. Of course, on their own, such diagnostic investigations do not necessarily identify the specific source or complete cause of a mismatch, but they can be used to compute sensitivities and, hence, point to likely suspects in the search for modelling errors.

At the global level, FSM fidelity can be expressed in terms of the relationships between the forces and moments acting at the aircraft centre of gravity and the resultant evolution of aircraft motion states. For small perturbations, the relationships are traditionally quantified in the form of 6-DoFs stability and control derivatives, or more generally, N-DoF generalised force and moment derivatives (e.g. including rotor dynamic and aerodynamics states). Such global analysis is also possible using test data, creating phenomenological models (p-models) e.g. using System Identification techniques to support fidelity assessment. As discussed earlier in this Chapter and later in the dynamic stability case study in Chap. 14, component-based force and moment derivatives can be used as a guide to FSM updating through fidelity renovation.

An FSM contains a myriad of computational pathways from interconnected local components to the global flight behaviour in the form of TSR predictions. Along these pathways, throughout the process network, data flows that provide the source of information that can be used in uncertainty analysis. Virtual (diagnostic) sensors within the network allow the user, or computer associate, to interrogate very detailed behaviour to establish where processes are relative to the boundaries of the DoP and DoV.

The sources of modelling deficiencies lie at the local level, propagating errors along the various pathways to the global level, e.g. in the magnitudes and directions of local flowfield, or in the way that component forces and moments react to the flowfield. Another example might be the way that a control actuation system saturates because of demands from the pilot/autopilot on the one side and/or loads induced by rotorblade torsion on the other. Some of the usual suspects of modelling deficiencies are described in Table 6.1. The list, including likely type of uncertainty, is far from exhaustive and can be modified or added to as experience is gained with the application of the RCbS process.

Once the primary source(s) of FSM error have been identified, the question becomes how to repair or renovate the deficiencies in fidelity. That brings us to the model tuning/updating process in Fig. 6.1.

Table 6.1 Some of the 'Usual Suspects' contributing to mismatches between simulation and test

Usual suspects	EUn or AUn	Issues and impact
Blade torsion dynamics	EUn	• Offsets between blade section c.g., a.c., t.c. give rise to couplings between flap, lag and (particularly) pitch/torsion • Cambered sections have pitching moment as function of velocity (dynamic pressure) as well as incidence and M, causing torsion dynamics even in steady flight • Change in 0/rev twist due to aerodynamic moment can require more or less collective in trim; impacts power • 2/rev Coriolis components in loads evidence for this effect
Aircraft inertias	EUn but can be treated as AUn	• Usually not measured so estimated from mass distributions; considered to be unreliable, particularly for old types where no digital data are available
Stick to blade calibrations	EUn but may vary from blade to blade and aircraft to aircraft so aspects of AUn also	• Calibrations usually made on the ground with auxiliary power unit (so actuation hydraulic pressure can be different from in-flight) • Feedback of rotor torsion aerodynamic loads through pitch-links to swashplate in flight will distort relationships
δ_3 Flap-pitch coupling	EUn	• Sometimes deliberately designed in (especially on tail rotor) but needs careful checking during calibration, so need flap measurements to be 'certain'

(continued)

Table 6.1 (continued)

Usual suspects	EUn or AUn	Issues and impact
Wake decay and contraction	EUn	• One of the most challenging aeromechanics prediction problems • Typical finite-state dynamic wake model has several 'tuning' parameters to shape results but these need to be physically realistic • Major impact on interference effects
Hub retention—flap, lag and torsion	EUn	• Modelling an elastic blade with a rigid blade, the hub retention structure is very important and dominates the hub moment predictions • Validation with blade tip and hub moment measurements can reveal how accurate are the rigid blade approximations
Fuselage blockage on empennage, tail rotor and variations with incidence and sideslip	EUn	• Known to be a significant contribution to reducing dynamic stability in forward flight • Variations with sideslip and incidence significant and has a dynamic component (hysteresis) giving rise to strong nonlinearities
Radial distribution of rotor dynamic inflow	EUn but usually no measurements for validation	• Significant impact on performance since induced drag/power impacted by loss of bound circulation radially • Variations during manoeuvres can impact dynamic response, particularly off-axis (e.g. Manoeuvre Wake Distortion) • Strong impact in vortex ring state

(continued)

Table 6.1 (continued)

Usual suspects	EUn or AUn	Issues and impact
Measurements of V, α and β Conversion to u, v and w	AUn and EUn	• Problem in low speed but also higher speeds • Impact of rotor wake can be significant and won't be captured in, e.g., wind tunnel calibrations • V, α and β vary at different points of the aircraft but usually measured at a single point, ahead of the nose on a boom
Kinematic consistency	EUn and AUn	• Crucial to undertake analysis to ensure that the measurements of re-constructed states u, v, w, p, q, r and θ, f, ψ form a consistent set that satisfy the 6-DoFs equations • Scale factor and bias errors are common for accelerations, angular rates and air-data measurements and can vary from flight to flight
Main rotor wake interference on tail rotor	EUn	• Can have major impact on yaw control in low speed manoeuvres with certain wind azimuth conditions; difficult to predict correctly without some form of vortex wake model
Tail rotor vortex ring state (VRS)	EUn	• Evident with winds from port side (anti-clockwise main rotor); requires significant corrections to tail rotor dynamic inflow model to predict the adverse effects of VRS

(continued)

Table 6.1 (continued)

Usual suspects	EUn or AUn	Issues and impact
Dynamic stall	EUn	• Can impact the performance and flight stability particularly for flight at high Mach number (forward speed, altitude) • Complex phenomena that triggers local unsteady rotorblade lift and pitching moment changes; stall characteristics quite different from quasi-steady stall and involve hysteresis
Yawed flow and reversed flow effects	EUn	• Complex nonlinear 3-dimensional effects that require aero data tables for sections that include large incidence and sideslip • Can lead to changes in aircraft flight behaviour at higher airspeeds

Model tuning and updating. In pre-certification FSM development, it is not unusual for fidelity to be deemed insufficient for the intended RCbS ACR. In the event the validation against test data reveals unacceptable discrepancies, the first step should be to investigate and reveal the cause of the discrepancy and postulate physics-based updates to the FSM. The update process could include modifying the modelling assumptions and/or adding previously un-modelled dynamics. Both might require the gathering of additional experimental data as illustrated by the iteration with flight testing in Fig. 6.1. Another option is to tune the FSM parameters to achieve the required sufficiency. Every design parameter in the FSM will have a degree of uncertainty (u_{inp}) and, within this established measure of uncertainty, sensitivity analysis can reveal the limits for parameter modification, or tuning, to increase fidelity. In the process, care must be taken to keep all parameters within physically meaningful bounds and to ensure that the aircraft-level tuning does not deteriorate the correlation against component-level test data. In case of doubt, it may be necessary to explore the limits of validity of a given parameter by comparison against a higher fidelity simulation approach. It is noted that, when accounting for simulation model validation uncertainty, u_{val}, the DoFs for improving the correlation with test data through input parameter variations (i.e. tuning) are naturally exercised and finding the unique combination of parameters that provides the 'best' match should always follow the physics-based doctrine.

If a system-level p-model is used, then all possible model-updating techniques could be applied, keeping in mind that such models can be used only for P1 predictability levels, i.e. interpolation only, since its DoR cannot extend, by definition, beyond the DoV.

A wide range of model-updating methods has been explored and documented by the NATO Research Task Group AVT-296 [6]; approaches applicable across the life-cycle of a rotorcraft, including for use in improving FSM fidelity in design, for training simulators and for certification. The physics-based update methods (methods 3–6 in [6]) are generally applicable to the certification process if applied within appropriate limits. AVT-296 also documented a variety of fidelity metrics commonly used as a part of the update process. Many of the methods described in the NATO report make use of System Identification methods to create p-models, derived from flight test measurements. Comparison of the parameters identified in the p-model structure with the equivalent parameters extracted from the FSM then provide the basis for the fidelity improvement activity. Example parameters are the 6-DoFs stability and control derivatives, rotor inflow deformation parameters, or constants in the rotor-blade retention modelling. Clearly, as with FSM parameter tuning, any adjustments to the FSM parameters based on such comparisons must be fully justified.

Finally, if the simulation model is used to obtain proof of compliance in the DoE, the credibility of the simulation model and validity of the underlying tuning must be explained and demonstrated. Credibility assessment is the topic of Chap. 9 in this Book.

6.4 Summary: Phase 2a

To summarise, activities in Phase 2a, development of the FSM based on the (preliminary) Requirements Specification, include:

i. FSM build, verification and validation and fidelity assessment, including updating/tuning and impact of uncertainties,
ii. Prototype FSM supplied to the FS and FTMS developers to support parallel activities,
iii. Input of pre-certification ground and flight test programme to support validation of FSM,
iv. Multiple iterative pathways managed and exercised as required throughout Phase 2a,
v. Updates to Requirements Specification based on Phase 2a activities,
vi. Outputs; FSM validated for intended purpose, documented in fidelity assessment reports.

References

1. Anon (2021) EASA concept paper: first usable guidance for level 1 machine learning applications A deliverable of the EASA AI roadmap. EASA
2. Anon (2012) CS-FSTD(H) certification specifications for helicopter flight simulator training devices. EASA
3. Mitchell DG, He C, Strope K (2006) Determination of maximum unnoticeable added dynamics. In: AIAA atmospheric flight mechanics conference and exhibit, Keystone, Colorado, USA
4. Padfield GD (2018) Helicopter flight dynamics, including a treatment of tiltrotor aircraft, 3rd edn. Wiley
5. Anon (2000) Aeronautical design standard performance specification handling qualities requirements for military rotorcraft. ADS-33E-PRF, United States Army Aviation and Missile Command
6. Tischler MB, White MD, et al (2021) AVT-296 rotorcraft flight simulation model fidelity improvement and assessment STO-TR-AVT-296-UU. NATO STO
7. Bryan GH (1911) Stability in aviation. MacMillan, London
8. Oberkampf WL, Trucano TG (2008) Verification and validation benchmarks. Nucl Eng Des 238(3):716–743. https://doi.org/10.1016/j.nucengdes.2007.02.048
9. DuVal R, He C (2018) Validation of the FLIGHTLAB virtual engineering toolset. Aeronautical J 122(1250):519–555. https://doi.org/10.1017/aer.2018.122
10. ASME (2009) Standard for verification and validation in computational fluid dynamics and heat transfer. The American Society of Mechanical Engineers, New York, NY
11. Roy CJ, Oberkampf WL (2011) A comprehensive framework for verification, validation and uncertainty quantification in scientific computing. Comput Methods Appl Mech Eng 200:2131–2144
12. Mehta UB, Eklund DR, Romero VJ, Pearce JA (2016). Simulation credibility: advances in verification, validation, and uncertainty quantification. NASA/TP—2016–219422
13. Perfect P, White MD, Padfield GD, Gubbels AW (2013) Rotorcraft simulation fidelity: new methods for quantification and assessment. Aeronautical J 117(1189):235–282. https://doi.org/10.1017/S0001924000007983

Chapter 7
Flight Simulator Development (Phase 2b)

7.1 Introduction

The focus of the guidance in this Chapter is on real-time, piloted simulations. The FS is intended to create an illusion of reality for the crew, so that they behave—are stimulated, react, and perform—as if they were in the real aircraft. Many factors contribute to this illusion. The fidelity of the various simulator features are obvious contributors, but also the protocols around how tests are conducted can reinforce or spoil the illusion. The test team must 'pretend' that they are conducting a real flight test and, as far as possible, engage in communications as if it was 'for real'. Even with a perfect FSM fidelity, the pilot's reactions, for example to failures, will depend on the cueing fidelity; failure identification, control strategy reactions and closed-loop recovery strategies need to be realistic. Achieving this kind of realism is no easy task, but rather calls for a development and validation discipline matching that described for the FSM. With the pilot-in-the-loop, the term 'behavioural fidelity' is also considered appropriate.

Figure 7.1 illustrates the elements of the FS Development phase (Phase 2b), with major inputs from the Requirements Capture/Build phase and the Engineering Design Data. Inputs from parallel Phases 2a (FSM) and 2c (FTMS) are also shown. The dashed lines indicate iteration pathways within Phase 2b and from outside, in the parallel phases and the later credibility phase. The steps required in the FS Development are:

1. Flight Simulator Build
2. Flight Simulator Verification
3. Flight Simulator Validation, including fidelity assessment.

It is acknowledged that there may be legacy FS facilities available to the applicant. In this case, the FS Build phase may be minimal, and the V&V activities may rely, in part, on past efforts, given appropriate configuration management practices are

G.D. Padfield et al., *Rotorcraft Certification by Simulation and Analysis*, Springer Aerospace Technology, https://doi.org/10.1007/978-3-031-86398-1_7

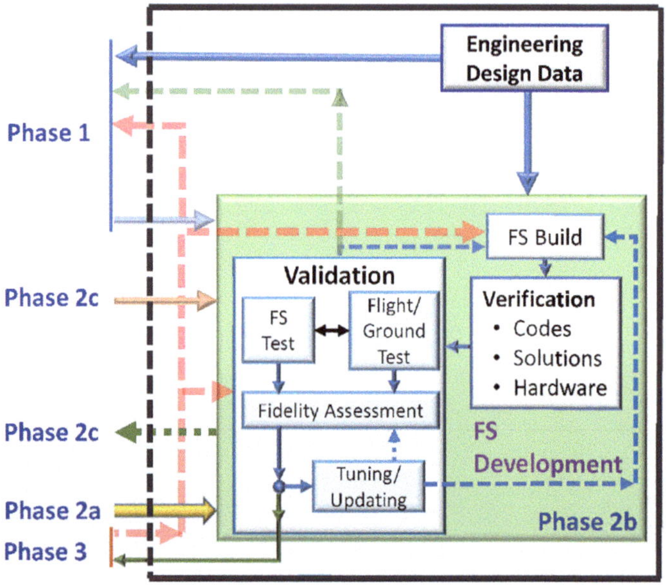

Fig. 7.1 The flight simulator development; Phase 2b

in place. The remainder of the activities within Phase 2b will then focus on the assessment and, if needed, the updating of the FS for the selected ACR.

An FS is comprised of different features (Fig. 7.2) which provide cues to the Evaluation Pilot (EP) enabling them to undertake Flight Test Manoeuvres (FTMs) relevant to an ACR. This Book draws on the definitions in Notice of Proposed Amendment—"*Update of the flight simulation training device requirements*" [1] and ICAO 9625 [2] for ten FS features. The FSM 'feature' developed in Phase 2a is a key component of the FS development as it provides inputs to the other FS features, e.g., the vestibular Motion Cueing System (VeMCS), and receives outputs from other features, e.g., the flight control positions and forces. The FSM content was discussed in detail in Chap. 6 and only the inputs/outputs to other FS features are discussed in this Chapter. For some ACRs and FTMs, not all FS features will be required, and these will be addressed in the following sections.

The Operator Station (OS) is the outer region of the FS schematic and interacts with the FSM and other features. The FSM provides inputs to the other FS features to generate cues to the EP at the centre of the FS schematic. The cues that can be provided to the EP are visual (sense of sight), auditory (sense of hearing), vestibular (sense of balance and orientation in space), proprioceptive and kinaesthetic (awareness of position and movement of joints respectively), and tactile (sense of touch). Each FS feature may generate one or more of these types of cues.

During fidelity assessment, predicted and perceptual fidelity metrics, defined in Phase 1, will be employed. The fidelity of the FS features should be sufficient to provide an EP with the cues needed to fly the FTMs associated with an ACR

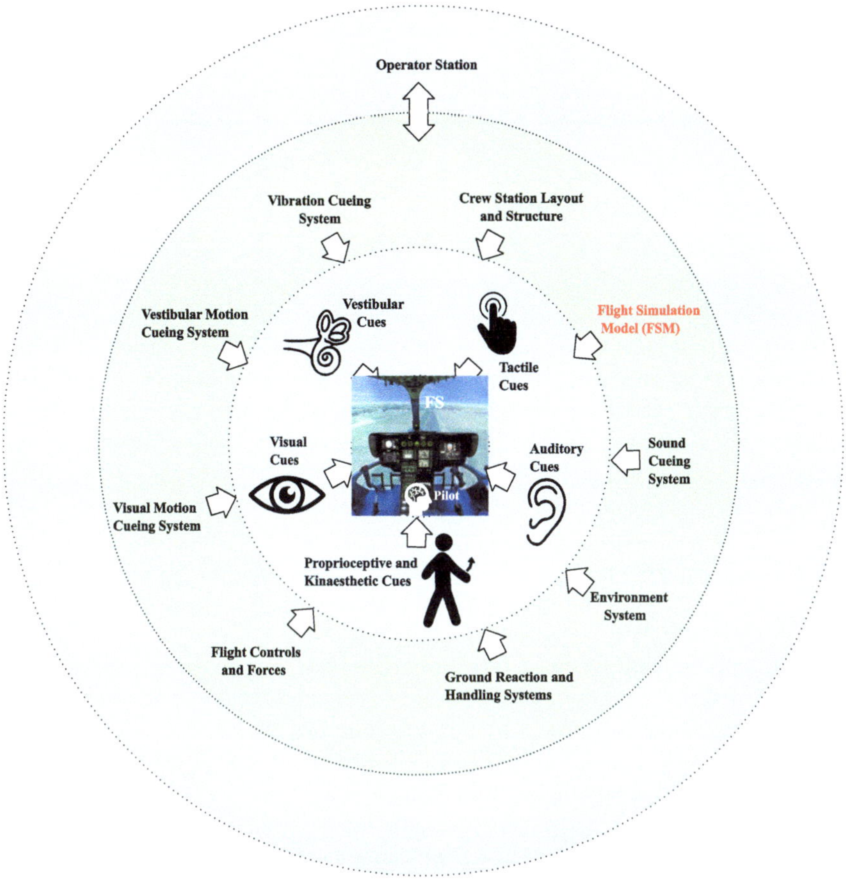

Fig. 7.2 Schematic of FS features

without 'significant' adaptation of control strategy and compensation. An example of a perceptual (behavioural) fidelity metric is the Simulation Fidelity Rating (SFR) scale [3]. The SFR scale (see Fig. 11.3) requires the EP to compare the task performance achieved and the task strategy adaptation required between flight and simulation to assess the fidelity of the FS. Quantifying task performance is normally straightforward, but strategy is more complex. An example of an adaptation metric, based on changes in control compensation is described in Refs. [4, 5]. The SFR method differs from current flight simulator training guidance [6] which defines Levels of Fidelity, e.g. specific, generic, and representative, to clearly differentiate certification guidance and allow more nuance in feature requirements. For reference, the certification-version of the SFR descriptors in Fig. 11.3, are suggested but not fully tested at the time of publication of this Book.

To emphasise, the content of the FS features should be such that they provide cues with sufficient fidelity for the FTMs associated with an ACR. Examples of the FS feature-content information are provided in the following sub-sections.

7.2 Flight Simulator Build

The design and build activities of the FS are based on the requirements specification for the selected ACRs. To repeat, it is crucial to ensure all FS features have sufficient fidelity to generate the cues needed by an EP to undertake ACR testing with representative task performance and without significant adaptation in control strategy. Cue fidelity is further facilitated through a fidelity assessment in the validation step of FS development. The guidance in this Chapter discusses the kinds of feature content that might be considered for an ACR.

The FSM is considered to be one of the features of the FS, but since its development and fidelity assessment are undertaken in parallel in Phase 2a (Chap. 6), it is not detailed here. A 'prototype' FSM, supplied by Phase 2a, can be used in the FS development to examine the sufficiency of cues such as the vestibular motion cues provided by the VeMCS.

An FS may be used for more than one ACR, so it is necessary to consider the fidelity requirements of multiple ACRs during the build step of the FS development process. Engineering design data from the real aircraft, e.g. aircraft cockpit Field of View (FoV) details, control forces/displacements, are also required during the FS build, to ensure sufficient fidelity of the FS cues.

It is recommended that applicants consider the design requirements for each FS feature in a structured way as detailed in Chap. 4 (Phase 1), where the following characteristics are considered:

1. Function
2. Operational modes
3. Data structures
4. Inputs
5. Outputs
6. Interfaces
7. Constraints
8. Domains of prediction and validation

It is acknowledged that the domains of prediction and validation become somewhat nebulous when considering a FS feature compared with the FSM. Consider the following example; piloted FS validation has been performed for a specific ACR resulting in a certain range of motion inputs from the VeMCS, thereby defining the DoV for the VeMCS. It is then found during the compliance demonstration trials in the FS that motion amplitudes/rates are larger, i.e., conceptually, the DoP might exceed the DoV. In this case, the applicant will need to show that the fidelity of the cueing from the VeMCS is sufficient within the full DoP, e.g. by validation at FS feature level to extend the DoV.

The following sub-sections define the nine FS features (not including the FSM) and discuss feature-content considerations for an ACR. The data structures and communication protocols for sharing information across features must be defined during the FS Build process.

Transmission of data between simulation features, and the time taken for a feature to cycle the data through a computation step, will introduce time delays in the FS environment, namely latency and transport delays. Latency is defined in CS-FSTD(H) [6] as the difference in the time taken for a real aircraft's system to respond and that taken for an FS feature to respond. The time taken for an FS feature to process an input signal from the EP's controls is defined as a transport delay. Transport delays will affect any solution produced by an FS feature, or collection of features, and must be quantified in the verification/validation of the FS solutions. Recommendations for maximum transport delay values, also considered applicable for RCbS, can be found in documents such as CS-FSTD(H).

7.3 Operator Station

Definition: The Operator Station (OS) is comprised of the computer terminal(s) from which the FS features are initialised, managed and monitored by the operator. The OS feature is not defined in [1], but for a certification FS the OS has content that needs consideration.

Design Considerations: The OS must enable the certification team, including the EP, flight test engineer, and the engineers operating and monitoring the FS, to gather data from the FS to demonstrate that the aircraft performance and piloting requirements for the ACR have been achieved. To support this process, the FS OS can be utilised to trim the FSM with the pre-certification flight test conditions flown (e.g. airspeed, weight, altitude, environment conditions) and record data from the simulated task. For example, in demonstrating compliance with ACR 29.143 (Controllability and Manoeuvrability) the OS can be used to interact with the Environment System (ES) feature to load the required visual database, and with the FSM to trim the model in a ground referenced hover in the presence of 17 kts winds for a range of azimuths.

The OS can record data obtained from the FSM, e.g. aircraft states, and the Flight Controls and Forces feature, e.g. pilot inceptor positions, for use in FSM and FS validation, fidelity and credibility assessment. The OS could include a real-time visualisation of the simulated aircraft's performance and the EP's control activity to enable the certification team to assess whether performance criteria have been met, whether the test point quality satisfies the ACR or if it needs to be repeated.

7.4 Environment System

Definition: The Environment System (ES) feature represents the FS components needed for the ACR which are not part of the FSM, but are configured through the OS. These include ATC, navigation signals, atmosphere and weather effects on the visual conditions (e.g. fog), as well as surface features such as aerodromes and terrain/landscape.

Design Considerations: The required characteristics of the ES are dependent on the ACR. For example, in CS-29, Appendix B, for IFR flight, the 'Environment—Atmosphere and Weather' characteristic should be able to represent Instrument Meteorological Conditions in the Visual Motion Cueing System (VzMCS), including visual range and precipitation effects, which are controlled by the OS. When the ACR FTM includes navigation tasks, navigational data should be included to provide inputs to the instrument panel in the Helicopter Systems feature. The ES feature includes the visual terrain database which comprises visual models of the terrain and provides height above terrain inputs to the Ground Reaction and Handling System, Crew Station and Layout Structure features, as well as to the FSM. If required for an ACR, landing areas and ground markings should be modelled with a sufficient level of detail, texture and contrast.

7.5 Ground Reaction and Handling System

Definition: Helicopter ground reaction and (ground) handling responses in an FS which are derived from the FSM.

Design Considerations: Although the landing gear/undercarriage component is a part of the FSM, it is included here as a separate FS feature as the input/output relationship needed for its operation, e.g. input from the ES system feature to the FSM and vice versa, must be considered during the FS design and build. The ES system feature will provide height above terrain information as an input to the FSM to calculate the undercarriage deflections, friction and forces. The outputs from these calculations are conveyed to the EP through the VeMCS and the VzMCS to aid in identifying the ground state of the aircraft, e.g. skidding or rolling motions. The height above terrain information is also needed for aerodynamic ground effect modelling in the FSM, which in turn affects the ground handling simulation. This simulator feature is not required for any ACR that does not involve landing, taxiing, contact with the terrain surface, or In Ground Effect (IGE) operations.

7.6 Crew Station Layout and Structure

Definition: The Crew Station is defined as the area in an aircraft where the flight crew members work and where the EP's inceptors, flight instruments and interfaces to helicopter systems such as the engine controls, are located.

Design Considerations: The content of this feature includes flight instruments, caution and warning displays, secondary aircraft controls such as buttons and switches to power and electrical systems, and navigation and communication systems. Some elements of the Helicopter System, such as the Stability Control Augmentation System (SCAS) and autopilot, are components of the FSM, and appropriate interfaces in the FS will be required to enable the EP to provide inputs to the FSM.

The construction, positioning and configuration of the content of this feature should be such that they do not noticeably modify the task strategy employed by an EP while performing an ACR, e.g., change in visual cues due to crew station framing and seating position. This feature could be developed with the use of a physical replica of flight hardware, or via a virtual representation of the cockpit. Engineering design data are required to enable the fabrication of the cockpit, or the generation of a 3D model for virtual applications.

The crew station layout and structure feature design will depend on the configuration of the VzMCS. A Virtual Reality (VR) VzMCS will require a digital recreation of the crew station, rather than a physical one. This may reduce the required complexity of the crew station significantly, but increase the complexity of the VzMCS in turn.

All aircraft systems required for an ACR should be available and appropriately located in the Crew Station, especially for tasks that include emergency procedures. It should enable any interactions required by the EP during the ACR with relevant aircraft systems. All systems required for accomplishing an ACR should be operable by the flight crew with no input from the simulator OS required.

The Human Factors aspects in Certification, although not specifically addressed in this Book are clearly relevant to the design of the Crew Station used in the FS. Relevant detailed features that are used in the flying tasks need to be sufficiently representative of the real cockpit that task strategies do not need more than minimal adaptation compared with flight.

7.7 Flight Controls and Forces

Definition: This feature is defined as the physical control inceptors used by the EP in the Crew Station and their force, displacement and damping characteristics.

Design Considerations: The FS inceptors can be broadly divided between 'unloaded' and 'loaded' (also known as 'force-feedback') systems. The former is typically associated with desktop simulation and will generally not represent the control ranges and forces that a pilot would experience in the aircraft. As such, their use would not normally be expected in the RCbS use of a FS. Loaded systems can allow customisable mechanical characteristics of the inceptors, e.g. breakout force and spring force gradient, to be generated to represent those of the aircraft. The requirements for sufficiency of the fidelity of these characteristics will be dependent on the ACR being assessed. The characteristics of the inceptors are typically provided by a control loading system, including computer and motor, and can be configured through the OS.

The EP should be able to apply control inputs in terms of amplitudes, forces and frequencies in a manner that is required for the FTMs associated with the ACR and that satisfy defined fidelity requirements. The output from the movement of the FS inceptors are inputs to the FSM's FCS component which, in turn, provides inputs to the FSM rotor components and SCAS/autopilot.

7.8 Visual Motion Cueing System

Definition: The VzMCS is any type of display technology, including dome projection and Virtual, augmented and mixed reality (VR/AR/MR), that provides out-the-window (OTW) visual motion cues that the EP uses for an ACR.

Design Considerations: Development of the visual cueing system must consider the required FoV, Field of Regard (FoR) (the total area that can be captured by a movable sensor), and appropriate levels of lighting and contrast to provide the EP with the visual flow information to attempt an FTM. For example, to enable the FTMs associated with ACR 29.143 (Controllability and Manoeuvrability) to be undertaken, the FoV and display resolution should provide sufficient visual motion cues to enable the EP to perceive similar height above ground and vehicle drift cues to that experienced in real flight.

For a projection system, the design eye-point is required from the Engineering Design Data to locate the pilot's head in the FS. For multi-channel projection systems, blending and warping of the visual channels on the projection screen/dome should not introduce any perceivable distortion in the OTW visual scene.

Although, at the time of writing, VR systems are considered immature for use in certification, they have been certified for use in rotorcraft simulator training [7]. It is anticipated that suitable options for creating the OTW visuals using VR, AR or MR systems will be commercially available in the future, and a brief description is included here. In VR, all visual cues are provided to the EP through a headset that generates the OTW cues and, possibly, a virtual cockpit. In this case, the EP does not see the physical simulator cockpit structure, nor their limbs and hands. Unless hand tracking and hand visualisation is included, this solution can only be used in ACRs where EP interaction with cockpit content other than the main inceptors (collective, cyclic and pedals) is not required, e.g. button presses. In the AR case, the EP can see the Crew Station Layout and Structure through the AR headset, while the OTW view is 'augmented' using a 'black mask'. Here, the Crew Station windows need to be blanked with black material or the room around the Crew Station needs to be dark or heavily dimmed. In the MR case, the real-time video image from two cameras mounted on the headset (one per eye) is mixed with a virtual representation of the OTW view (using a mask of the Crew Station or making use of chroma-key techniques). In the MR case, specific aspects of the video pass-through need to be considered, such as time delay of the video image, degradation of the image quality and the distortion caused by the offset between the pilot's eye location and the camera position in front of them.

With AR/VR/MR devices the following aspects need to be considered when choosing the hardware. The vision of the pilot is limited by the FoV of the headset, while in simulators with a projection system the OTW FoV is limited by the projection system. On the other hand, the FoR of head-mounted devices (AR/VR/MR) is $\pm 180°$, whereas the FoR of a projection system is limited by the physical structure of the cockpit. The minimum FoV/FoR required for an ACR demands specific consideration. Too small a vertical FoV can lead to adaptation of the EP's eye scanning and increased head movements and discomfort. The latency between tracking of head movement and image rendering can cause 'cybersickness'. Furthermore, reference [8] suggests that having a frame rate less than 90 frames per second can cause misperception of motion, also leading to cybersickness. A mismatch of the distance between the centres of the EP's eyes, the inter-pupillary distance, and that set in the hardware can cause fatigue and cybersickness. For the MR case, care needs to be taken with the camera position to prevent parallax issues and false motion perception during head rotation. Further guidance related to reducing cybersickness in VR is provided in [8].

7.9 Sound Cueing System

Definition: The sound cueing system is any technology that provides auditory cues to the EP.

Design Considerations: In flight, the EP receives sound cues from several sources, providing feedback on the helicopter's state, e.g. engine/rotorspeed, gearbox whine and indications such as audio cautions and warnings, e.g. low rotorspeed warning. The inputs to sound cueing system are provided by the FSM and other features, such as ATC communications from the ES. The presence of unrealistic, or incoherent, sound content can impact an EP's situational awareness and sense of realism, and the sufficiency of the sound cue feature in the FS must be considered for each ACR. Particularly in failure cases where the pilot's attention is focused on the outside world visual cues, such as during a Category A rejected take-off, sound cues are likely to have an impact on task performance. Sound cueing systems should also include intercom facilities that allow the EP to communicate with other crew members and the FS operator. Tones and voices from the avionic system are a source of information for the pilot, and, in several situations, being able to 'pick-up' rotorspeed changes from the rotor/engine/gearbox tones will be important.

7.10 Vestibular Motion Cueing System

Definition: The VeMCS provides rotational/translational displacement, rate and acceleration cues to the EP.

Design Considerations: The human vestibular system contains two important motion sensors, located in the inner ear: the semi-circular canals for rotational cues, and the otoliths for translational cues. The semi-circular canals are sensitive to angular velocities and accelerations, while the otoliths sense translational motion, particularly surge, sway and heave accelerations. The VeMCS produces the vestibular cues through the Motion Drive Algorithms (MDA), or motion filters. By tuning the parameters of the MDA, the fidelity of the vestibular cues can be optimised with the need to keep the movable cabin of the VeMCS within its available motion space (see text box).

A wide range of VeMCSs may be employed that can vary in the number of the Degrees of Freedom (DoFs) they provide, the maximum velocities and accelerations they can produce, and the envelope of the system, i.e. the maximum translational or rotational displacements achievable by the moving platform. The VeMCS is driven with the translational accelerations and angular rates calculated by the FSM which are scaled and filtered, using the MDA, to keep the resulting platform motion within its operating envelope. The MDA filters have coefficients that can be used to tune the response of the VeMCS.

Careful considerations of the MDA filter coefficients, the gains and break-frequencies, are needed to ensure the EP receives sufficient vestibular cues for the FTM of an ACR. The VeMCS should not provide any adverse cues as perceived by the EP. Ideally an MDA should use a high enough gain to provide sufficient onset cues, whilst also having a low phase shift to reduce adverse cues. Sinacori [9] proposed a set of predictive VeMCS fidelity levels using gain and phase criteria. Later, Schroeder [10] reported on simulation experiments that supported revisions to Sinacori's criteria. Use of a high gain will result in more of the motion platform envelope being utilised, potentially reaching a limit, and will also require more motion washout, resulting in a larger phase shift between the flight-model output and the motion platform response, to return the platform to the platform's neutral position. The motion platform washout should occur at a level that is not 'intrusive' to the EP and that keeps the phase shift at an 'acceptable' level. More recently, the Objective Motion Cueing Test (OMCT) [2] has been proposed as a predictive fidelity metric. It compares the calculated motion response of a flight model (motion system input) to the actual movement of the VeMCS (output) as modified by the MDA. The MDA coefficients will need to be tuned for each motion platform DoF or combination of DoFs, according to the ACR, to provide sufficient positive cues to an EP. An example of a process that could be used to determine MDA filter coefficients is provided in [11] where it was shown that sufficient vestibular motion cues could be achieved by careful harmonisation of the 2-DoF motion filter gains for a roll-sway task, outside of the criteria defined by Sinacori and Schroeder.

Considerations in the tuning of the mdas with a small-medium motion system

The tuning parameters within the motion drive laws can be optimised for individual test manoeuvres. Failure to do this introduces the risk that adverse motion cues might reduce perceived fidelity below sufficiency level. Aspects that need to be considered in this tuning process include:

a. Envelope of expected motion in a flight task about each axis in terms of accelerations and rates,
b. Distinguish between primary and secondary motion/control axes; this supports optimising the distribution of MDA commands to platform 'legs',
c. How might translational-rotational motion compensation be achieved to correct for adverse cueing in a task, e.g. roll-sway or pitch/surge,
d. Define expected ranges of pilot 'gains' in the task. This can be useful in the tuning to optimise for medium gain levels (e.g. moderate level aggression),
e. Ensure the matching of transport delays in the visual and vestibular motion systems.

7.11 Vibration Cueing Systems

Definition: Vibration cueing systems are any technology that provides proprioceptive, tactile and vestibular motion cues to the EP in the frequency and amplitude range associated with pilot seat vibrations.

Design Considerations: The motion frequency range that a VeMCS can produce is normally limited (0–10 Hz in the heave axis and lower in the surge and sway axes [6]). High-frequency VeMCS motion can put additional unwanted loading on other FS features such as the VzMCS. To generate higher-frequency vestibular motion cues that may be associated with, e.g. retreating blade stall in high-speed flight or vortex interactions in the translational lift phase, a separate Vibration Cueing System, e.g. in the form of a vibration seat or platform, could be included in the FS. The DoFs in the Vibration Cueing System will be dependent on the FTM designed for the ACR.

7.12 Flight Simulator Verification

The codes, solutions and hardware of each FS feature must be verified during the FS development process, i.e. the construction, functionality and operation of each feature must be consistent with the requirements of the FS specified in Phase 1. Each feature will have inputs to/from other features in the FS, e.g. the FSM provides inputs to the VeMCS to generate vestibular motion cues, the ES and the Ground

Reaction and Handling System both have inputs to the FSM to produce ground contact responses using height above terrain information. The applicant needs to demonstrate the requirements specified in Phase 1 have been correctly realised during the FS verification assessment.

Codes. The FS features will have associated computer codes which must be verified. The applicant should ensure that any codes that operate on an input correctly produce the expected output. This process identifies errors that might occur due to, for example, inconsistency of units used, rounding errors in calculations and definitions of axis reference systems. The applicant needs to demonstrate that they have an appropriate configuration management process (see Chap. 3) in place so that the effect of any updates to codes are verified before implementation into an FS feature. Codes are typically compiled into a real-time version and the data generated by the real-time version should be compared with the data generated by the source code to ensure they agree.

An example of the verification process is provided as follows. The FS requirements related to testing for an ACR might dictate the use of a motion platform to provide vestibular motion cueing for the EP. Outputs from the FSM are inputs to the MDA, which in turn drives the motion platform actuators. A simulation model of the motion platform response to actuator demands could be independently developed and the response of the FS VeCMS compared for verification. This could be achieved by inputting step or sinusoidal signals into the MDA and comparing the response with the input demand. This approach has been proposed in the OMCT [2].

Solutions. Most FS code is deterministic. However, with an FS having multiple interacting features, using different time clocks to produce solutions, this may result in non-deterministic behaviours, or behaviour that is very difficult to predict. A check is required to ensure that these solutions are as expected and permit execution in real-time.

For example, the ES provides height above terrain information to the FSM. Contact between the undercarriage component of the FSM and the terrain surface must be detected and then trigger a series of calculations to compute forces and moments on the undercarriage and the resulting vehicle motions. The output from the FSM is then transmitted to other features such as the VzMCS, VeMCS and the sound cueing systems. These complex interactions must be verified to produce consistent solutions, if these effects are required for the ACR being assessed.

Within a single feature, there may also be 'variability' in solutions based on inputs. For example, the graphics engine in the ES will require computation time to produce a solution following an input from the FSM, i.e. an update of the aircraft inertial position. The computation time, and the resulting framerate of the graphics signal produced, may vary based on the detail in the visual scene, for example, the number of polygons used to represent macro-textures, and the resolution of micro-textures. The framerate should be maintained above the minimum requirements defined in Phase 1 of the RCbS process, informed, for example, by values contained in documents such as CS-FSTD(H) [6].

Hardware. The construction and function of any FS feature hardware must be verified to match the requirement specification for that feature. This process may include the following:

1. Checking physical dimensions and movement ranges of inceptors against technical drawings,
2. Checking that 'input' functions of hardware devices drive the intended mechanical or computational response, e.g. a button press generates an on/off signal as required,
3. Checking that 'output' functions of hardware devices respond to commands as intended, such as speakers playing a sound, without unexpected distortion, at an appropriate volume.

Relevant delays that exist in the real-world system, e.g. pressure instrument lag, should be appropriately modelled in the FSM, or associated FS feature, and verified in the FS. The output from an FS feature such as the displacement of an EP's inceptor, should be verified against the value received and used by the FSM FCS component.

If Phase 1 (FS requirements) specifies the use of a VeMCS for an ACR, its response to input commands should be verified. An example of this process is given in ICAO 9625 in the form of the OMCT [2]. The OMCT defines a matrix of translational and rotational inputs, of varying amplitude and frequency, to evaluate the ability of the VeMCS to reproduce these commands. The resulting platform motion can be verified using motion sensors, e.g. accelerometer measurements.

In case of an existing FS, part of the hardware verification can be derived from documentation from past activities with the requirement that configuration management has been fully exercised.

7.13 Flight Simulator Validation

The FS validation process is intended to ensure that the cues that the FS features generate are of sufficient fidelity (representative) to enable the EP to undertake the FTMs of an ACR realistically, i.e. effectively equivalent to flight. At its heart, the sufficiency assessment is, therefore, a comparison between task performance achieved and control strategy employed in the FS and the real aircraft. The validation process illustrated in Fig. 7.1 is divided into three iterative steps:

1. Testing
    a. FS Test
    b. Flight/Ground Test
2. Fidelity Assessment
3. Tuning & Updating

The testing step is divided between flight/ground test of the aircraft, and testing using the FS. These steps can take place in parallel, but it is expected that they will be informed by a common strategy. The following sections will address each of these steps in more detail.

It is assumed at this stage that the FSM is sufficiently 'mature' to conduct FS validation. As mentioned previously, a prototype FSM can be used during the FS development. However, it is recognised that the validation processes may take place in parallel to some degree, and the FS development can also support the FSM validation process through subjective evaluation if FSM tuning/updating is required.

FS test. FS testing is used to assess the features of the simulator and the cues that they produce to enable step 2 of the FS validation process, the fidelity assessment. It is expected that applicants will use a structured approach to planning a FS test campaign to design repeatable, performance-bounded tasks relevant to an ACR. This starts by defining the aim of the testing, descriptions of the manoeuvres to be flown, aircraft test configuration, and the data to be gathered. In Phase 1 of the RCbS process, objective and subjective metrics will have been defined to assess FS fidelity, which are then adopted in the FS test campaign.

A process for designing flying tasks is provided in ADS-33 [12], where performance requirements for Mission Task Elements (MTEs) are defined based on the role of the aircraft. It is suggested here that a similar approach is adopted to develop performance requirements for FTMs associated with an ACR.

It is recommended that a FS test trial document is developed and reviewed by the test team prior to testing. It is suggested that EPs are selected that have flight experience of the ACR, and on the aircraft to be certified, to provide 'informed' feedback during the validation process. It is expected that a trial briefing is completed prior to any testing so that it is clear to the test team what tasks will be flown and that any clarifications on test requirements are addressed, e.g. task performance, control strategy.

The task and aircraft configuration are controlled using the FS OS feature. During testing, the EP can be reminded of the task description and performance requirements prior to undertaking a test point. Feedback can be provided to the EP regarding the task performance achieved using data recorded by the OS. It is recommended that EP feedback is sought and recorded on all elements of the FS test campaign, in support of validation.

A subsequent de-brief session enables the test team to review the results of the tasks flown, the EP feedback including any ratings awarded, and to provide supporting data for use in the Credibility Assessment, Phase 3. A trial report should be produced to collate this information.

Flight/Ground test. Data should be gathered from flight/ground testing, within the DoV, to support FS fidelity assessment. It is suggested that tests are designed to enable comparisons with FS testing, where it is safe to do so. The aircraft should be instrumented, as defined in the Requirements Specification, to provide data for both FSM and FS fidelity assessments; more information on the FTMS is provided in Chap. 8. Data for FS fidelity assessment could include measurements of control activity, or subjective handling qualities ratings for MTEs/FTMs related to the ACR.

A similar process to testing on the aircraft can be used. However, the ground testing procedure may differ if a pilot is not required. For example, for some aircraft the mechanical characteristics of the flight controls can be validated via ground test measurements. However, for aircraft with changing mechanical characteristics over the flight envelope (e.g. due to force feedback through pitch-link rods), additional measurements during flight might be required.

Fidelity assessment. The fidelity assessment process evaluates whether the FS fidelity is sufficient for the relevant ACRs and selected I-P Levels. It uses outputs from the FS and flight/ground testing to compare objective and subjective metrics. If fidelity of the FS is assessed as sufficient, then the applicant can proceed to the Credibility Assessment phase. If FS fidelity is lacking after assessment, then a 'tuning/updating' process can be undertaken to correct identified deficiencies for use in future iterations of the FS validation process.

The fidelity assessment process is informed by objective and subjective metrics and associated tolerance margins, defined in Phase 1 of the RCbS process. Unless deficiencies are encountered, and tuning/updating of a specific FS feature is required, the FS fidelity assessment, in principle, only considers the full aircraft and the interactions with the pilot. The related metrics fall into two categories:

1. Task performance
2. Task strategy

For objective measures, a comparison of the task performance achieved in flight and simulation cover parameters specified in the task definition for FS and flight/ground test. For example, the FTM developed for CS 29.62 Rejected take-off: Category A, may have performance requirements on the touchdown conditions of the aircraft, which can be directly compared between FS and FTMS data.

Task strategy can be quantified through analysis of the pilot's control activity. Biometric measures can also be used to inform such comparisons. For example, pilot control strategy can be quantified by computing control input displacement/rate using the so-called attack activity [13] or other time- and frequency-domain metrics [5]. Biometric measures may include head and eye tracking data, heart rate, or brain activity (electroencephalography), and an applicant would need to demonstrate that these measurements indicate sufficiency of the FS fidelity for an ACR in terms of comparative pilot compensation, FS vs Flight Test (FT).

Subjective methods should also be used during the fidelity assessment to capture the EP's experience. These methods rely on feedback from the EP on the sufficiency of the FS fidelity for an ACR. This feedback can be sought through, e.g., the use of subjective rating scales, such as the Cooper-Harper Handling Qualities Rating (HQR) scale [14], or the Bedford workload rating scale [15]. These scales allow an EP to assess the level of performance achieved and the associated compensation required (HQR scale) or the spare capacity for a task. Comparisons of such ratings, awarded in FT and FS testing, allow a fidelity assessment to be made. In addition, it is recommended that the SFR scale [3], as described in Chap. 11, is used in the fidelity assessment process as it was specifically developed to provide a framework for FS fidelity evaluations. Whilst the SFR scale was initially developed for application to training tasks, there is a direct read-across for RCbS as described in Chap. 11. Use of the SFR scale addresses two aspects:

1. Comparisons between flight and simulation task performance, i.e. the precision with which a task is completed, and,
2. Task strategy adaptation, i.e. the degree to which the pilot is required to modify their behaviour (change the form of compensation) when transferring from simulator to flight and vice versa.

The robustness of the awarded SFR will depend on the pilot's ability to reflect on their task strategy and the achieved task performance. Therefore, the pilot must be proficient in flying both the vehicle and the task, and must also have operational currency so that a meaningful fidelity assessment can be made. Caution is advised during the fidelity assessment process, as a pilot can quickly adapt to FS deficiencies. Capturing this adaptation via pilot reflection needs to take this into account. A fidelity questionnaire is useful in the assessment process to identify any FS deficiencies.

FS tuning/updating. The source of any perceived FS fidelity deficiency needs to be unambiguously traced to the related FS feature to support FS tuning/updating. For example, the FS fidelity requirements defined in Phase 1 might dictate the use of a specific feature e.g. a VeMCS, and during the fidelity assessment the EP might perceive false 'motion' cueing. This can arise due to VeMCS content deficiencies, e.g. in the motion filter gain and break frequency, but also due to deficiencies in the FSM, or the VzMCS. In some cases, subjective fidelity assessment of the related feature(s) can be conducted using an appropriate rating scale, such as a motion fidelity rating scale [16], to aid in identifying and correcting the feature content. It is recognised that tracing the source of FS fidelity deficiencies can be challenging, but the guidance in this Book stresses the importance of facing such challenges purposefully.

The results of any tuning of the FSM would need to be assessed in the FSM development phase (2a) prior to re-evaluation in the FS fidelity assessment process.

During the RoCS project, several flight simulators were used to explore different aspects of the RCbS process [17, 18]. A substantial exercise of the process was undertaken for three ACRs on the University of Liverpool's HELIFLIGHT-R simulation facility [19] in July 2023. The results of these 'Case Studies' are reported in [20–22], with the main results documented in Chaps. 12–14. The HELIFLIGHT-R facility and the various rating scales and in-cockpit-questionnaire used during the simulation trials are documented in Chap. 11.

7.14 Summary: Phase 2b

Summarising the activities in Phase 2b, development of the FS based on the (preliminary) Requirements Specification:

i. Prototype FSM supplied to the FS to support parallel activities,
ii. FS build, verification and validation and fidelity assessment,
iii. Draw on pre-certification flight test programme to support validation of the FS,
iv. Multiple iterative pathways managed and exercised as required throughout Phase 2b,

v. Updates to Requirements Specification based on Phase 2b activities,
vi. Outputs; FS validated for intended purpose (for the selected ACRs), included in fidelity assessment reports.

References

1. Anon (2020) NPA 2020-15 Update of the flight simulation training device requirements. EASA
2. Anon (2012) ICAO 9625 Manual of criteria for the qualification of flight simulator training devices. ICAO
3. Perfect P, Timson E, White MD, Padfield GD, Erdos R, Gubbels AW (2014) A rating scale for the subjective assessment of simulation fidelity. Aeronautical J 117(1189):953–974. https://doi.org/10.1017/S0001924000009635
4. Memon WA, Cameron N, White MD, Padfield GD, Lu L (2021) The development of a pilot control compensation metric for simulation perceptual fidelity assessment. In: 47th European Rotorcraft Forum, Glasgow
5. Memon WA, White MD, Padfield GD, Cameron N, Lu L (2021) Helicopter handling qualities: a study in pilot control compensation. Aeronautical J 126(1295):152–186. https://doi.org/10.1017/aer.2021.87
6. Anon (2012) CS-FSTD(H) Certification specifications for helicopter flight simulator training devices. EASA
7. Anon (2022) EASA approves the first virtual reality (VR) based flight simulation training device. Retrieved from https://www.easa.europa.eu/newsroom-and-events/press-releases/easa-approves-first-virtual-reality-vr-based-flight-simulation.
8. Proietti P et al (2021) Guidelines for mitigating cybersickness in virtual reality systems TR-HFM-MSG-323. NATO STO
9. Sinacori JB (1977) The determination of some requirements for a helicopter flight research simulator. NASA-CR-152066
10. Schroeder JA (1999) Helicopter flight simulation motion platform requirements. NASA/TP-1999-208766
11. Hodge SJ, Perfect P, Padfield GD, White MD (2015) Optimising the roll-sway motion cues available from a short stroke hexapod motion platform. Aeronautical J 119(1211):23–44. https://doi.org/10.1017/S000192400001023X
12. Anon (2000) Aeronautical Design Standard-33 Performance Specification (ADS-33E PRF): handling qualities requirements for military rotorcraft. US Army Aviation Engineering Directorate
13. Padfield GD, Jones JP, Charlton MT, Howell S, Bradley R (1994) Where does the workload go when pilots attack manoeuvres?—an analysis of results from flying qualities theory and experiment. In: 20th European Rotorcraft Forum, Amsterdam
14. Cooper GE, Harper RP (1969) The use of pilot ratings in the evaluation of aircraft handling qualities. NASA TN D-5153
15. Roscoe AH, Ellis GA (1990) Subjective ratings scale for assessing pilot workload in flight: a decade of practical use. Royal Aircraft Establishment RAE TR 90019
16. Hodge SJ, Perfect P, Padfield GD, White MD (2015) Optimising the yaw cues available from a short stroke hexapod motion system. Aeronautical J 119(1211):1–22. https://doi.org/10.1017/S0001924000010228
17. Duda H, Gerlach T, Advani S (2013) Design of the DLR AVES research flight simulator. In: AIAA Modeling and Simulation Technologies (MST) conference, Boston, MA, 19–22 August
18. NLR (n.d.) Helicopter pilot station. Retrieved from https://www.nlr.org/research-infrastructure/helicopter-pilot-station-hps/

19. White MD, Perfect P, Padfield GD, Gubbels AW, Berryman AC (2011) Acceptance testing and commissioning of a flight simulator for rotorcraft simulation fidelity research. Proc Inst Mech Eng Part G J Aircraft Eng 227(4):663–686. https://doi.org/10.1177/0954410012439816
20. White MD, Dadswell CM, Lu L, Padfield GD, Quaranta G, van't Hoff S, Podzus P (2023) Case studies to illustrate the rotorcraft certification by simulation process; CS 29/27 CAT A. In: 49th European Rotorcraft Forum, Bückeburg, Germany, 5–7th September
21. van 't Hoff S, White MD, Dadswell CM, Lu L, Padfield GD, Quaranta G, Podzus P (2023) Case studies to illustrate the rotorcraft certification by simulation process; CS 29/27 low-speed controllability. In: 49th European Rotorcraft Forum, Bückeburg, Germany, 5–7 September
22. Lu L, Padfield GD, White M, Quaranta G, van 't Hoff S, Podzus P (2023) Case studies to illustrate the rotorcraft certification by simulation process; CS 29/27 dynamic stability requirements. In: 49th European Rotorcraft Forum, Bückeburg, Germany, 5–7 September

Chapter 8
Flight Test Measurements for FSM/FS Development (Phase 2c)

8.1 Introduction

Critical to the success in the validation process of both the FSM and FS is the quality of the flight test measurements used in the comparisons between reality and simulation. To set the scene, we recall a famous adage, of unknown origin (although something similar has been attributed to Albert Einstein), related to the development and validation of rotorcraft simulation models.

> *No one believes the simulation result, except the person who created it,*
> *while everyone believes the flight test data except the person who measured it.*

The message here is that, while there will always be doubts about the fidelity of simulation, and while we expect flight test measurements to be the 'truth', those who have been closely involved in designing, building and using an FTMS, understand the meaning of the second part of the adage. They are aware of the significant number of pitfalls involved in the design and build of an FTMS, and they understand the importance of care and attention to detail required to mitigate or avoid these pitfalls and to minimise measurement uncertainty.

The increased use of M&S in support of certification will require not only increases in FSM/FS fidelity but also a sustained emphasis on the quality of the test data used in pre-certification validation activity. Another important point is that the achievement of the envisaged levels of fidelity and credibility are likely to require in-flight measurements from the rotor systems, to support validation and understanding of physics-based model updates. To emphasise this point, it is no exaggeration to state that the lack of rotor measurements (e.g. pitch, flap and lag angles) can sometimes make it very difficult, if not impossible, to determine the cause of flight behaviour poorly predicted by the FSM. Blade flapping, lagging and pitch/torsion measurements can provide critical information for the engineer to chase down a behavioural deficiency and aid in the (physics-based) tuning of e.g. parameters in the rotor inflow or the flap-pitch (δ_3) coupling models. Consideration can also be given to using a

G.D. Padfield et al., *Rotorcraft Certification by Simulation and Analysis*, Springer Aerospace Technology, https://doi.org/10.1007/978-3-031-86398-1_8

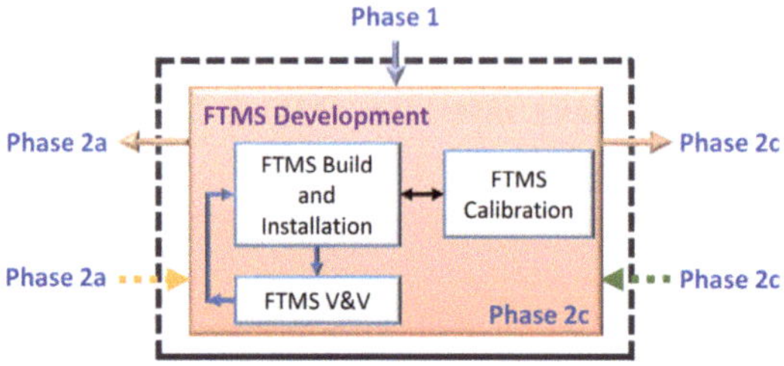

Fig. 8.1 Phase 2c in the RCbS process; FTMS development

test aircraft instrumented for rotor structural loads substantiation to support the FSM development.

Figure 8.1 shows the sub-phase from Fig. 2.1. In this Chapter, we describe some of the important issues to be addressed when designing, building, calibrating, installing and using a FTMS, including the extraction of data from the system and its use by the FSM/FS engineers. The five topics will be addressed separately, noting that at each stage, quality can be preserved or degraded through the decision making and practice of the FTMS engineers or shortcomings in the initial FTMS requirements specification. This Chapter ends with a discussion on the expected content of a pre-certification Flight Test Guide (FTG), noting that the certification FTG will also be augmented by companion testing with the FSM and FS.

The V&V for the FTMS follows a similar process to the FS. In a general sense, the integrated H/W + S/W FTMS must be proven to meet the documented requirements (Verification) and must operate such that it provides the intended outputs (Validation), i.e. the requirements were correct. The calibration process (Chapter 8.4) is critical in this regard as it is where physical quantities of interest (e.g. accelerations, aerodynamic velocities, rotor flap motion) are sensed and converted into electronic information for comparison with 'validated' benchmarks, e.g. instrumented inertial platform. The FTMS is the source of validation data for both the FSM and FS and S/W that converts raw measurements into the information required for the validation must pass through a V&V process.

8.2 FTMS Design

As emphasised for the creation of the FSM, the design specification of the FTMS should be based on requirements. The requirements must address the measurement functions, their precision, resolution and range, allowable levels of measurement and process noise, methods of calibration and installation, and the process of data capture; including sampling rates, synchronization, any relevant analogue-to-digital signal conversion, associated filtering and the interface of the FTMS with the crew and ground station. The design specification should also include requirements relating to the building of the FTMS. Measurement redundancy should be taken advantage of in the design of the system and associated data processing (e.g., velocities from air data, inertial data and satellite navigation data). The requirements set for a comprehensive rotorcraft FTMS will be extensive and need to be developed in close collaboration with the developers of the FSM itself. Depending on the FSM requirements set in Phase 1 and the validation activities foreseen, it may be necessary to include a model of the FTMS (a virtual prototype), as part of the FSM, featuring sensor locations, updatable calibrations and data processing. The same argument applies for embodying a prototype FSM within the FTMS development of course. A plan for verifying and validating the design should be included as part of the design (e.g. as in Ref. [1]).

8.3 FTMS Build

The requirements for the building of a FTMS should emphasise the integrated nature of the system, maintenance requirements and usability; and the purpose of the system, which will include the acquisition of data for use by flight simulation experts in model validation. The FTMS will be built as an integrated set of sub-systems; e.g. the air data, the inertial data, flight controls, rotor flap and lag dynamics, rotor loads, engine/transmission, satellite navigation and so on. There will likely be a requirement to include measurements used by any control augmentation system, e.g. stability, autopilot, load alleviation, as well as flight information available to the crew, e.g. presented by a glass-cockpit system, noting the challenges involved in accessing data from proprietary systems. The integration process should ensure that the FSM expert is presented with a coherent, consistent set of data, digitized to the same sampling rate. A common clock for all sub-systems is normally required to ensure measurements can be related coherently but, even then, time shifts in sampling will result in time shifts in time histories. The maximum time shift between two time histories is likely to be related to the minimum sampling rate. For example, if blade pitch, flap and lag angles are required every 5° of rotor azimuth to capture dynamic stall effects, a sample rate of several hundred Hz will be required; the same for the tail rotor might increase by a factor of 5. In contrast, an adequate sampling rate for dynamic pressure, sideslip and incidence, normally measured on a boom attached to the front fuselage, might be 100 Hz. The variable sampling rates mean it is important to consider the post-measurement data integration processing as part of the FTMS. The need for data in the 'rotating system' should be addressed in the design specification, including how these are transmitted to the recording system (e.g. through slip-rings or radio transmission), and synchronised with non-rotating data.

8.4 Calibration

Calibration requirements fall into two categories; off-board and on-board. For example, inertial measurement systems are commonly calibrated using an off-board motion table, while rotorblade flap angles usually require on-board calibration. If two different sources of calibration are available, it is advisable to compare the two, and quantify the levels and characteristics of the measurement and process noise present in the on-board data. This particularly applies to air data measurements using a boom. For example, a pitot-static and vane system can be calibrated in a wind tunnel for comparison with on-board calibration data. The latter can be derived from flights in still air involving tracking a ground pace vehicle, with the aircraft trimmed at various pitch/incidence and heading/sideslip angles. The wind-tunnel calibrations will need to take account of the process noise due to tunnel wall effects. The on-board system measurements will include the effects of rotor wake impingement on the boom, a source of process noise that is difficult to quantify. At low speed, the level of such process noise might be so large that confidence in the measurements becomes too low for use in validation. If low airspeed measurements, and related sideslip and incidence, are critical to the fidelity level required, then an appropriate sensor system must be used. Calibration of controls should consider the pathway from the pilot station, through the powered actuation system, the swashplate and control linkages to individual blade pitch angles; a set of measurements that may have different static and dynamic behaviours. Accurate measurement calibration is clearly critical to the successful use of a FSM in certification.

8.5 Installation

The locations and attachment methods used for the FTMS and its sub-systems are also important. For example, fuselage motion sensors are best located close to a nominal centre of mass, with translational motion sensors, e.g. accelerometers, isolated from translational vibration and the anti-nodes of structural bending, while rotational motion sensors, e.g. rate gyros, isolated from rotational vibration and the nodes of structural bending. Requirements for the installation of sensors that capture control motions and rotor system behaviour should address the acceptable levels of 'intrusion' to preserve the integrity of the measurements. The interfaces of the data capture system with the crew and ground stations are important for real-time monitoring and review of data quality, involving the installation of dedicated telemetry and cockpit display systems. These aspects are further considered in the next section.

8.6 Usage, Including Flight Testing

The stage is set as they say; the FTMS design, build and calibration is complete and the system, having met the design requirements, is installed in the aircraft and on-board calibrations are completed. The productive part of the process now begins with the FTMS usage and FT campaign. As already emphasised, pre-certification flight trials to validate the FSM will take on a new level of importance as they gradually replace the certification trials themselves. The pre-certification flight test campaign will be defined in a comprehensive trial plan, including aircraft configurations to be tested, coverage of the flight envelope and, critically, crew instructions for the characterisation of aircraft behaviour in terms of trim, stability and response in defined weather conditions. The test campaign should involve close coordination with the development of the FSM and FS. Effectively, test points flown in the real aircraft can be pre-tested with the FSM, either off-line or in a piloted simulation environment, as appropriate. Comparisons between the FSM and FT steer the identification of FSM flaws and the assessment of potential updates. This integration of the FT campaign with the FSM development can have a major impact on progress, but is so important, to the avoidance of nugatory testing on the one hand, and to facilitate efficient model-updating on the other, that it should be embraced as a fundamental aspect of pre-certification flight trials. The need, and scope, for innovation in this area is significant. In this respect, it is recommended that a more detailed and generic FTG be developed in a collaboration between certification agencies and applicants. In advance of this, the following sub-section discusses key aspects of the content of such a FTG.

Pre-certification flight test guide. As with all activities within the RCbS process, the pre-certification FTG must be based on, and reflect, the requirements of the FSM/FS validation and fidelity assessment processes. The close-coupling between requirements for the FTMS with those for the FSM and FS is reinforced within the content of the FTG and documented within the context of the DoV relevant to the ACR under consideration. For example, in the 4-dimensional matrix of conditions for quantifying static and dynamic stability (airspeed, density altitude, c.g. location and aircraft mass), 70% might be selected within the DoV (30% in the DoE). Of these DoV points, 60% might be selected for RCbS. With this plan, across the whole of the DoP (DoV + DoE), only 28% of points would be flight tested in Certification. In the pre-certification testing, a subset of such points might also be selected for fidelity assessment.

In support of FSM validation and fidelity assessment, linking with the content of Chap. 6, a typical set of test points at each flight condition/aircraft configuration might include:

(a) Step/pulse control inputs to exercise the air data measurement system across its anticipated range of variation. This provides the core data for the so-called kinematic consistency analysis intended to ensure that air data system and inertial data system measurements of aircraft velocities form a consistent set. Any calibration error corrections derived from this analysis will need to be applied to data derived during sorties to capture data for fidelity metrics in the DoV.

(b) Trims across the required ranges of incidence and sideslip to provide the core data for quantifying static stability in forward flight or controllability in low-speed manoeuvres.

(c) Control frequency sweeps (one control axis in turn) that provide the core data for system identification analysis to create p-models for use in the fidelity assessment and updating of the FSM. Real-time, or post-run, analysis to check the coherency across the frequency range of interest should be conducted and the sweeps repeated until input–output coherences, and hence the test data, meet the defined quality requirements.

(d) Multi-step control inputs (one control at a time) to provide the core data, for example for frequency and damping of the lateral-directional oscillation to derive its stability characteristics using pedal control doublets. Another example might be related to the RTO, recording responses to collective and cyclic control in steep descent conditions. Data from, for example, 2311-type multi-steps can be used in the 'validation' of the p-models before they are used to support the FSM fidelity assessment.

(e) Two repeats of test points for all the above are advisable to support the resolution of any anomalies.

In addition, flight test points in support of the FS validation and fidelity assessment will be required, linking with the content of Chap. 7. These could include:

(f) Tests to exercise failure modes, e.g. SCAS lane failures. In addition to capturing data for the FSM (response) fidelity assessment, such tests would enable the pilot to assess the various failure cues, including vestibular motion, in support of the FS fidelity assessment.

(g) Characterising the HQ deficiencies resulting from AFCS failures.

(h) FTM testing to support the FS fidelity assessment for ACRs that involve task flying, e.g. CAT A RTO, controllability and manoeuvrability in cross-wind hover. Such examples could involve the use of the SFR scale to enable evaluation pilots to quantify the sufficiency of the FS cueing.

It is suggested that the FTG be written by the RCbS engineer in close collaboration with the evaluation pilot(s) and FTMS engineering team. Agreed criteria for test point quality and success/failure need to be defined and in-flight judgements and decisions made, based on such criteria. Evaluation team training in the application of (open-loop) test inputs can be conducted in the FS to ensure maximum data quality, e.g. duration and steadiness of trim prior to control input, input magnitude and shaping, input and response duration, criteria for recovery.

It is recognised that the Phase 2 'pre-certification' testing could imply that additional flight and simulation testing will be required in the RCbS process, conducted at the selected DoV points. In the Phase 3, Certification, testing, some points could be the same as those already flown in the pre-certification tests, while others will form the matrix of points 'within' which interpolation will be performed using the FSM/

FS. The distribution of the pre-certification and certification points in the DoV will be established as discussed in the first paragraph to this sub-section and will depend on the Influence Level(s) selected for the ACR. The extent of the DoE will also be a factor in establishing the distribution of points in the DoV, since an understanding of the trends in fidelity and the extent of fidelity updating within the DoV will be factors in quantifying uncertainty and hence credibility of the DoE results.

The significance, and hence importance, of flight test data quality is highlighted here, recognising that the distribution of RCbS points across the DoP, although initially defined by the I-P matrix in Phase 1, will probably not be finalised until the fidelity assessment within the DoV is complete. The FTG should be written to accommodate this flexibility.

Finally, to emphasise, uncertainties in the test data used in fidelity assessment must be quantified by the FTMS engineers, based on the analysis of calibrations, kinematic consistency, measurement and process noise assessments. Quantification of u_r needs to be undertaken here and be a key output of Phase 2c.

8.7 Summary: Phase 2c

To summarise, activities in Phase 2c, development of the FTMS based on the (preliminary) Requirements Specification, include:

i. FTMS design, build, calibration and installation,
ii. Alignment between FSM/FS validation needs and FTMS,
iii. Create the pre-certification Flight Test Guide,
iv. Execution of pre-certification ground and flight test programme to support validation of FSM,
v. Multiple iterative pathways managed and exercised as required throughout Phase 2c,
vi. Quantification of referent uncertainty, u_r, and what is required to meet the minimum acceptable requirements defined in Phase 1,
vii. Updates to Requirements Specification based on Phase 2c activities,
viii. Outputs; FTMS and validation test data for Phase 2a and 2b.

Reference

1. Padfield GD (1991) AGARD-LS-178: SA 330 Puma identification results. AGARD

Chapter 9
Credibility Assessment and Certification Activity (Phase 3)

9.1 Introduction

With Phase 2 and the initial fidelity and uncertainty assessments of both the FSM and FS complete, the applicant moves into Phase 3, Credibility Assessment and Certification Activities (Fig. 9.1). Within this phase are the defining moments for the achievement of certification, so it is expected that Certification Authorities will be even more closely involved. Credibility addresses the complete set of RCbS results for the chosen ACRs across all selected I-P Levels (Table 4.2).

Demonstrating Credibility within the DoV, including the results of interpolation, is anticipated to be relatively straightforward, and rooted in the Phase 2 results of fidelity assessment and FSM/FS updating/tuning, leading to validation. Also, the uncertainty analysis conducted in the DoV can give the model developer a scale to assess the credibility of the DoE results, noting that when the validation error is comparable with the uncertainty, the model is likely within the precision achievable given the data and software available (see Chap. 6.3).

To emphasise, results within both the DoV and the DoE will need credibility evaluations in Phase 3 before the case for Certification can be sufficiently well evidenced and this Chapter discusses how such evidence may be assembled and presented.

Several general kinds of extrapolation can be considered. The first, typically in Predictability Levels 2 and 3 (Table 4.2), involves cases where the extrapolations consist of extensions of fidelity assessments made within the DoV, e.g. based on a validated model with proven physics-based updates. Extrapolating assessments made at low-altitude into the high-altitude regime could be an example here. Another might be the extrapolation of level flight dynamic stability to climbing/descending or turning flight.

The second kind of extrapolation, typically in Predictability Level 4 (Table 4.2), involves cases where the ACR being considered is not supported by directly comparable results in the DoV, e.g. landing following total power loss. But even in such cases, there are likely to be fidelity analyses that can be drawn on from the

G.D. Padfield et al., *Rotorcraft Certification by Simulation and Analysis*, Springer Aerospace Technology, https://doi.org/10.1007/978-3-031-86398-1_9

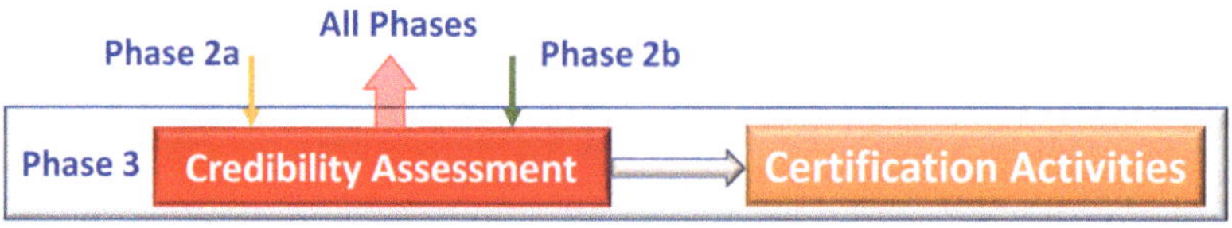

Fig. 9.1 Entering Phase 3, credibility assessment and certification activities

DoV that inform fidelity assessment and credibility in the DoE, e.g. results from autorotation flight tests conducted at altitude, including entry and recovery.

A third, and special, kind of extrapolation could relate to STC applications, i.e. changes to an original approved design. e.g. hoist or external installations. While this Book does not feature examples or case studies in this category, it is envisaged that the proposed RCbS process will still need to be followed, and founded on the original certification. If initiated by a third party, close collaboration with the OEM will be required to enable such STC applications. More generally, having established an RCbS basis for a rotorcraft, the application of the process for any such life-cycle developments by the OEM is likely to be very efficient.

Three considerations are suggested to maximise confidence and the credibility of these, and other, kinds of extrapolations:

(a) Develop an extrapolation from several points within the DoV,
(b) Understand, through analysis, the physical sources of variation in predictions in the DoV (e.g. of performance margins or fidelity deficiencies),
(c) Understand, through analysis, how these physical sources may change in the DoE and what other kinds of physical sources might need to be considered.

For all kinds of extrapolation, applicants need to describe the location within the DoE, relative to the boundaries of the DoV and the DoP.

The credibility of the results from extrapolation must be informed by a thorough uncertainty quantification (UQ), a topic touched in Chaps. 4 and 6, and expanded on in this Chapter.

As with fidelity, the notion of sufficiency is also important for Credibility. To emphasise the point, at the RCbS process point reached at entry to Phase 3, to achieve sufficient simulation Credibility, it is considered essential that;

(1) A thorough verification and validation has been conducted to identify and address, to the extent possible, the various sources of prediction error that might influence prediction accuracy.
(2) Fidelity within the DoV has been quantified and errors and uncertainties for the FSM/FS predictions and FTMS outputs are characterised.
(3) Extrapolations into the DoE are informed by the three considerations, (a), (b) and (c) described above, as well as past-experience from the applicant regarding the evolution of simulation model error and uncertainty along the extrapolation dimensions.

(4) The development and exploitation of the simulation framework has been performed by a team with the necessary expertise and experience following a controlled development process, akin to what is proposed in this Book.

The fourth point has already been highlighted on several occasions in this Book and its importance will feature strongly as the results of the Credibility assessment are presented to the certification authority.

The various sources of test data error and prediction error addressed in the (Phase 2) validation process have been discussed in Chaps. 6–8. The current Chapter deals with how believable the results of simulation are, their Credibility, along with the evidential basis, simulation uncertainty and its characterisation. As recommended throughout this Book, it is assumed that the applicant has employed physics-based modelling and updating for the FSM, exercised throughout the DoR.

It is important to re-emphasise at this point that the V&V processes within the Phase 2 developments are intended to ensure that the FMS, FS and FTMS all meet the requirements defined in Phase 1; particularly that the functions and operations within the three elements meet the defined fidelity sufficiency standards. In this way, applicants are expected to strengthen confidence in their ability to quantify uncertainty through the V&V and fidelity assessment processes, and hence quantify credibility in the predictions.

9.2 Simulation Credibility and Uncertainty

Simulation credibility assessment. In the current context, simulation credibility refers to the extent to which the predictions from the FSM or FS are faithful representations of reality, have sufficient fidelity and can be relied upon to assess the compliance of the aircraft to the selected ACRs, considering the potential uncertainties in the simulation and test data. The essential elements of a credibility assessment have been enumerated above, and previously in Chaps. 4 and 6. To re-iterate, the concept of simulation credibility is particularly (but not exclusively) relevant in a certification context where the DoP extends beyond the DoV into the DoE, a region where the simulation error cannot be fully assessed based on a comparison with test data, considering simulation error and uncertainty. Credibility in the DoE relies on the applicant's perception, ultimately shared by the certification authority, that they are addressing what are normally described as known unknowns, as well as having an awareness of the so-called unknown unknowns, that inform and underpin the quantification of confidence.

While a generally accepted framework for flight simulation credibility assessment currently does not exist, numerous efforts have been made in various fields of science and engineering [1–5]. There is also an ASCE/ME Journal of Risk and Uncertainty in Engineering Systems dedicated to this topic, that would be expected to report up-to-date research relating to mechanical and civil engineering. It is emphasised that the guidance laid out in this Book does not advocate any specific method. What is important from a regulatory perspective is that the essential elements of simulation credibility assessment are adequately addressed by the applicant, including UQ.

Uncertainty analysis and quantification. In the AIAA's recommended practice for using simulation to support certification [4], UQ is expressed in terms of four elements;

(a) Identification (Where are the major sources of uncertainty?),
(b) Characterization (What form are they, and what are their mathematical descriptions?),
(c) Propagation and Aggregation (How do they combine to determine total uncertainty in the analysis results?), and
(d) Analysis (What are their impacts and implications?)

Reference [4] discusses these four elements in detail and makes an important point relevant to the RCbS process; "*Community-wide adoption of addressing analysis uncertainty using the structure of these four elements will facilitate clear communication between applicants, regulatory authorities and other industry stakeholders.*" This Book therefore endorses this recommendation in pursuit of the same communication goal.

In Chap. 6, in addition to model-form uncertainties, the three elements of validation uncertainty were introduced: the uncertainties due to numerical errors u_{num}, those associated with referent (experimental) errors u_r, and those from errors in the input parameters u_{inp}. To be clear, the errors (δ's) and related uncertainties (u's) are defined as the effects on the simulation output, e.g. control/performance/stability margin, allowing them to be included directly in the fidelity and credibility assessments in Phases 2 and 3 with, for example, δ_{val}. The maximum acceptable uncertainties in the input parameters (e.g. moments of inertia), referent parameters (e.g. air data calibration scale factors) and numerical parameters (e.g. discretisation factors) need to be defined in Phase 1. At the input level, parameter X_i has an uncertainty $u_{Xi}^{(inp)}$ with output $u_{Xi}^{(out)}$.

Uncertainty in the input parameters should be part of the data provided in RCbS Phase 1 from the design department, supplemented with expert insights on the type of modelling included in the FSM. In some situations, an applicant might elect to include some of the model-form parameters, e.g. from within the rotor wake interference model, as part of the u_{inp} set, rather than including them as uncertainties stemming from model-form errors (u_{model}). Uncertainties then might reflect (epistemic) ranges over which such parameters preserve the physics of the modelling. In other situations, the uncertainty in such parameters might reflect the model-form errors. Both approaches are considered acceptable, but clearly need to be defined for all input parameters that have associated uncertainties, in Phase1. Assessing the impact of the input uncertainties on the simulation outputs will have been undertaken in Phase 2 as part of the FSM validation. However, it is likely that some knowledge of the relationship between input and output uncertainty will also have been available in Phase 1, e.g. from previous experience, to guide the requirements specification. Regarding flight test data, the effects of referent uncertainty should be determined within the Phase 2 FTMS development and calibrations (Chap. 8). Solution verification is the process by which u_{num} is estimated. To emphasise, the requirements for (maximum) acceptable uncertainties in all three categories will need to have been defined as part of the Requirements Specification in Phase 1.

Typical FSMs make use of parameters that are quantified through specific experiments or, in some cases, inferred from design requirements and data. In principle, all these data should have uncertainties associated with them that could either be of an epistemic nature, without a prescribed probability distribution, or aleatoric due to known random variations that exist from one aircraft or component to another or from time to time. In any case, given estimates of these data uncertainties, it is possible to estimate the effect on the output quantity of interest. Examples of the approaches that might be used to obtain such estimates, derived from a priori estimates of the input parameter uncertainty, are as follows:

a. Local linear analyses e.g., using Taylor series expansions for the simulation result of interest, to determine the (linear) sensitivity coefficient derived from the FSM, due to the input parameter (X_i) uncertainty u_{Xi}. This approach leads to a local assessment, i.e. close to the values of the nominal parameters values.
b. A more general, global, statistical approach without assumptions of linearity, based on Monte Carlo or other similar stochastic methods. Typically, the required numerical effort is higher. In this case, the input parameter uncertainty is defined in the form of probability distributions. The numerical burden falls on the computer, of course, and it is the physical interpretation of the results that enables the user to draw meaningful conclusions.
c. If the parameter uncertainty is epistemic (EUn), interval analysis could be used, that may be less expensive in terms of computational burden compared with Monte Carlo or other statistical methods.

Input parameter uncertainties can sometimes be estimated from prior experiments, and, for an input X_i, u_{Xi} may be characterised not only in terms of an interval with unknown probability (epistemic uncertainty), but also in terms of a statistical distribution (aleatoric uncertainty). Epistemic uncertainties should not be represented by a uniform probability distribution, i.e., an interval within which all values are equally likely, because this assumption adds information and, unjustifiably, reduces the assigned uncertainty in the input data. For aleatoric uncertainties, the assumption of a Gaussian distribution requires, at a minimum, the definition of a mean and standard deviation. As discussed in Chap. 6, the confidence in a mean estimate of an (random, Gaussian) uncertain parameter can be quantified through the standard deviation and associated 'coverage' factor k; 95% confidence for $k = 2$ and 68% confidence for $k = 1$.

Rather than a specified uncertainty, one could assume the (conservative) worst-case parameter value (i.e. leading to a worst-case impact on simulation output), if indeed this limit can be justified (e.g., minimum specification engine power, or conservative control rigging). Such an assumption might be made if the uncertainty is epistemic, with an unknown probability across a range; the value at the extreme of the range being a worst case, e.g. c.g. location. The concept of conservatism is further

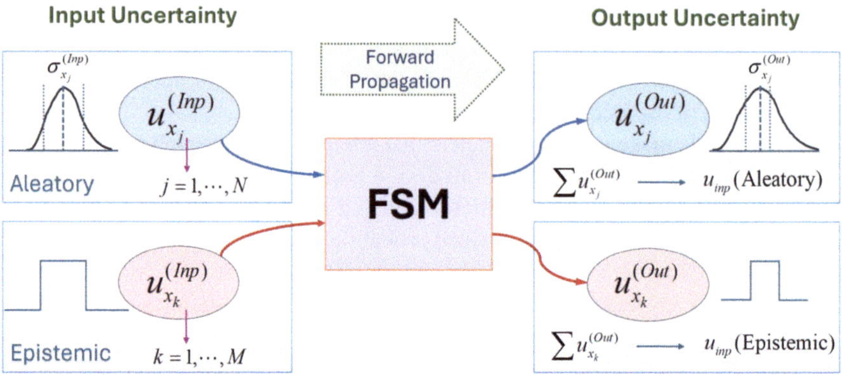

Fig. 9.2 Illustration of the propagation of the impact of aleatoric and epistemic uncertainties at the input level on the FSM outputs

discussed later in this Chapter. If statistical data are lacking, it is useful to seek expert opinions to gain, e.g. information on intervals. Even in their simplest forms, these approaches provide the applicant (and certification authority) with an understanding of the simulation output variability and primary parameters of influence which need to be characterised with a degree of certainty as high as possible.

As an illustration, Fig. 9.2 shows the process of propagation by which simulation output variability might be related to parameter input variability.

For aleatoric uncertainties, where the input parameter uncertainty u_{Xj}^{Inp} might be known in the form of a probability distribution, forward propagation through the FSM leads to a similar form of output distribution, with mean u_{Xj}^{Out}. For epistemic uncertainties u_{Xk}^{Inp}, known to be contained within a range, propagation produces an output range as shown. In this book, the parameter dressing are generally omitted for clarity, so that we refer to, e.g., u_{inp} whatever the nature of the input parameter uncertainty.

It is worth noting that the input parameter uncertainty analyses described above also extends to the virtual pilot (see Chap. 6). In this case, the analyses may be used to explore the effects of the uncertainty in piloting strategy, e.g. where deviations from the RFM specified procedures are deliberately introduced to verify the efficacy of the specified alternate procedures. In taking this route, the applicant must carefully consider the interrelationship between aircraft model input uncertainty and the uncertainty due to the virtual pilot model and its parameters.

For cases where the simulation is applied in the DoE, an assessment must be made of how the prediction errors and uncertainties from the DoV are expected to evolve along the extrapolation dimension. This assessment may be aided by, for example, access to historical (flight test and simulation) data, or comparisons with higher-order numerical prediction methods. The extrapolation limits specified in the FAA's AC 29.45 [6] have been established based on historical flight test experience and contemporary analytical and simulation methods; e.g. "*a predicted controllability model developed for high altitude may be used if verified by limited flight testing with steady ambient winds. The extrapolation guidelines in AC 29.45 b(2) are still applicable. These high-altitude controllability tests could typically be conducted in conjunction with take-off, landing and performance tests.*" and "*Controllability can usually be extrapolated up to a maximum of 2000 feet above the highest test site altitude.*" It is noted that the FAA refer to 'verification by limited flight data', using the term verification rather than validation as used in this Book. As required in the FAA/ EASA standards, the present guidance advocates that applicants exercise informed engineering judgement to evaluate whether the defined limits for extrapolation are applicable, particularly for novel aircraft configurations and advanced analytical and simulation techniques.

To emphasise, it is essential, for the assessment of simulation credibility, that the uncertainty throughout the prediction domain (validation (interpolation) and extrapolation) is adequately characterised. Within this context, it is useful to again make the distinction between aleatory and epistemic uncertainty. Epistemic uncertainty (EUn) is present in the simulation if there is a specific lack of knowledge, or shortcomings in understanding, about the physical processes being modelled. Included in this definition are model-form uncertainties stemming from the approximate representation of reality, e.g., in the form of the rotor wake and interference modelling. In principle, EUns can be reduced by gathering additional test data or by using increased complexity, or higher-order, numerical tools such as Computation Fluid/Structural Dynamics to gain further insight into the physics being simulated. In the latter case, the knowledge gathered from the higher-order solution may also be used to reduce the uncertainty bounds on specific input parameters. However, residual EUns are likely to remain and need to be accounted for.

Aleatory uncertainty (AUn) is due to the (multiple) inherent variations in the physical system and its parameters and is probabilistic in nature. This type of uncertainty might include variations in aerofoil properties (due to manufacturing tolerances, erosion and repairs), vehicle moments of inertia, control riggings, atmospheric turbulence, etc. AUns can be addressed in several ways, typically requiring many simulations to be run. In fact, this is a clear benefit from the use of the structured RCbS process to replace or augment flight test, given that simulation provides an efficient means of evaluating a very large number of variations on a given scenario. The aleatoric nature of test data, represented by the referent uncertainty u_r, is familiar in validation analysis where so-called scatter is present in, e.g. air data and accelerometer measurements due to process noise.

AUns can, in principle, be addressed in a probabilistic manner using UQ techniques [4]. This approach relies on the identification and characterisation of the sources of uncertainty in the modelling, and aggregating and propagating these throughout the model to obtain output probability distributions. The benefits of this approach to UQ can only be fully realised when the capability to interpret such distributions, in the context of extrapolation, are well developed by the applicant.

Another approach to dealing with uncertainty (in an input parameter or at prediction level) can be described as conservatism. That is, by introducing conservative or (statistically) worst-case assumptions, and/or limitations in the FSM or FS cueing environment, leading e.g. to under-prediction of margins, or by imposing conservative limits of acceptability on the compliance parameters. In these cases, it must be shown that the assumptions are indeed worst-case and/or that the effect of the assumptions are large enough to account for the prediction uncertainty. The problem with this approach is, of course, that it can lead to suboptimal, or unnecessarily reduced, aircraft performance if used excessively. Furthermore, it needs to be shown that the conservatism does not invalidate the physical modelling. That is, the conservative assumptions must be physically meaningful such that the simulation does not deviate from the DoR. The approach of dealing with uncertainty through conservative assumptions is not uncommon to the AMCs, as implied in AMC25.1309 11.e.4c, which states that "*uncertainty should be accounted for in a way that does not compromise safety*".

Confidence ratio. Within the broader concept of simulation credibility, the predictive quality of the simulation hinges on the extent of both the simulation error and validation/predictive uncertainty. To emphasise, increasing the credibility of the prediction requires reducing both the error and the uncertainty, e.g. by FSM updating, or by reducing the uncertainties in the various input parameters.

The required level of credibility of the simulation prediction (initially stipulated in Phase 1, see Chap. 4.2) is tied to the proximity to non-compliance, i.e. the margin (M) between the prediction and the boundary of the performance requirement. Thus, the closer the case is to being non-compliant in a certification sense (smaller M), the lower must be the acceptable uncertainty U in the simulation or, more generally, in the components of the validation error, δ_{val}, (in the DoV) and prediction error, δ_p, (in the DoE). As discussed in Chaps. 4 and 6, this dependency can be captured using the concept of the Confidence Ratio, *CR*, illustrated again in Fig. 9.3, and generally defined in terms of the ratio of the 'distance' between the FSM prediction and the performance requirement, or the margin M, to the uncertainty U in the prediction.

$$CR = \frac{M}{U} \tag{9.1}$$

So, as the uncertainty U increases, confidence reduces. A 'large' *CR* implies either that the case is far from being non-compliant (large M), or that the combined uncertainties (U) in the simulation prediction are low compared to the distance to the performance requirement. In Fig. 9.3 the 'estimated referent/experimental value', and related (estimated) simulation validation error (δ_{val} in DoV) or prediction error (δ_p in DoE) have been added for reference. They are shown as non-conservative, but they could of course be conservative, with the blue line featuring to the right of the performance assessment, and the errors suggesting an increased margin (and *CR*) to the limit. The estimated referent value in the DoE might be derived from extrapolation of the test data in the DoV, although these can only be a rough guide to what might be measured. The estimated δ_p, strictly unknown in the DoE of course (only the simulation value S is computed), should be embedded in the uncertainty U, transforming to an effective uncertainty, U_e, as discussed in Chap. 6.

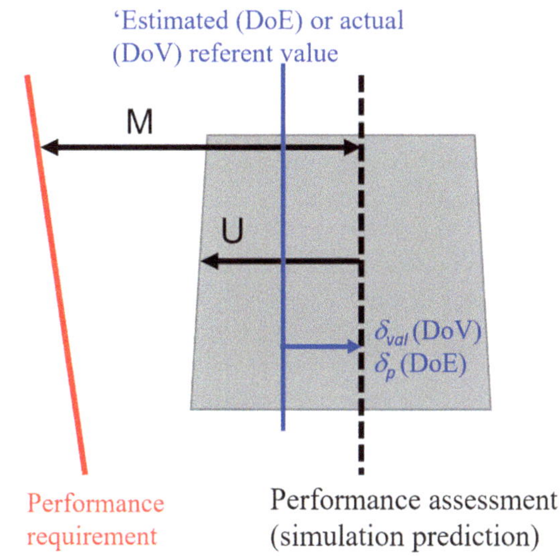

Fig. 9.3 Schematic illustration of Confidence Ratio Parameters, including the simulation validation/ prediction and associated errors δ_{val} and δ_p

As noted, the embedding can lead to increased or reduced *CR*, depending on whether the trend in the DoV was for under-prediction or over-prediction of the margin *M*. This will be returned to when discussing stability margins later in this Chapter.

The *CR* was first introduced in Chap. 4, since applicants must make their initial estimates of credibility and related uncertainties for the Phase 1 Requirements Specification. There, in Table 4.3, reproduced as Table 9.1, the relationships between the *CR* and an applicant's confidence in the results of simulation for a specific ACR were suggested. In turn, how this confidence translates into the RCbS activity for particular ACRs was illustrated in Fig. 4.4. The case was speculative, but reinforced the point already made that the closer the prediction is to being non-compliant, the higher should be the required confidence (i.e. the lower uncertainty), and therefore the credibility, in the conclusions drawn from simulation. The concept of 'closeness' is akin to tolerance relating to fidelity sufficiency; it is subjective, must be informed by expert knowledge and clearly related to the impact on safety. Even if it may prove infeasible to define *CR* requirements that hold universally true, the concept does provide an intuitive normalised quantification of the confidence in the simulation predictions.

Table 9.1 Suggested Confidence Ratio (CR) ranges

$1.0 < CR < 1.1$	Minimum Acceptable confidence (MA)
$1.1 \leq CR < 1.25$	Medium confidence (M)
$1.25 \leq CR < 1.4$	High confidence (H)
$1.4 \leq CR$	Very High confidence (VH)

To emphasise, the *CR* concept described above applies to those parameters for which a performance requirement, and therefore a margin, exists within the ACR, e.g., the control margin for a controllability assessment [7], or the damping of an oscillation for a dynamic stability assessment [8], and is particularly relevant, but not exclusive, to cases in the DoE. It is anticipated that requirements for the *CR* should be ACR-specific. Generally, a *CR* greater than unity is required to account for "unknown unknowns", even though these are likely to be very small. Finally, as extrapolation will increase the uncertainty in the predictions, this will automatically be reflected in a reduction of the *CR* for cases of extensive extrapolation which, by implication, are strongly discouraged in this guidance.

An example is now presented to provide insight into how the *CR* might be used in practice. The case is drawn from the RCbS Case Study reported in more detail in Chap. 12.

ACR 29.143(c) (Controllability and Manoeuvrability) requires that the "*wind velocities from zero to at least 31 km/h (17 knots), from all azimuths, must be established in which the rotorcraft can be operated without loss of control on or near the ground in any manoeuvre appropriate to the type.*" The applicant has proposed that simulation is used to compute the trim pedal margins up to the maximum take-off and landing altitude for the so-called critical azimuth (in this case with wind from approximately 90° starboard). The safe margin (i.e. the performance limit) is defined by the red line in Fig. 9.4 and has been established based on flight testing in the DoV, demonstrating that the indicated margin ensures adequate control authority. The applicant is seeking partial-credit for this ACR, limiting the use of simulation to the prediction of the control margin.

Figure 9.4 illustrates a possible set of results. The chart on the top left (a) presents the outcome of the validation analysis in the DoV, expressed as a function of the intended extrapolation dimension (in this case, density altitude). The chart shows the comparison between flight test and FSM, with the (uncertainty) bars generally indicating an interval, or probability bounds, depending on the nature of the uncertainties. The uncertainty for the simulation predictions in the DoV, as plotted, reflects the validation uncertainty, centred around the prediction mean. The estimated validation/prediction error, related to the range within which the model form error δ_{model} lies, ($\delta_{\mathrm{val}} \pm \mathrm{u}_{\mathrm{val}}$), is shown in the top right chart (b). The validation uncertainty $\mathrm{u}_{\mathrm{val}}$ (in the DoV) and prediction uncertainty u_{p} (in the DoE) are plotted in the bottom left chart (c), assuming a linear extrapolation of $\mathrm{u}_{\mathrm{val}}$ (which, implicitly, is also an extrapolation of u_{r}). The bottom right chart (d) then presents the result of subsequent application of the extrapolated model form uncertainty to the pedal prediction, embodying the uncertainty u_{p}, to derive the CR, as computed using the expression:

$$U = u_p = \left(\sqrt{u_{num}^2 + u_{inp}^2 + u_{model}^2}\right) \tag{9.2}$$

Here, $\mathrm{u}_{\mathrm{model}}$ is taken from Fig. 9.4c, and the numerical and input uncertainties are recomputed for the prediction condition in the DoE.

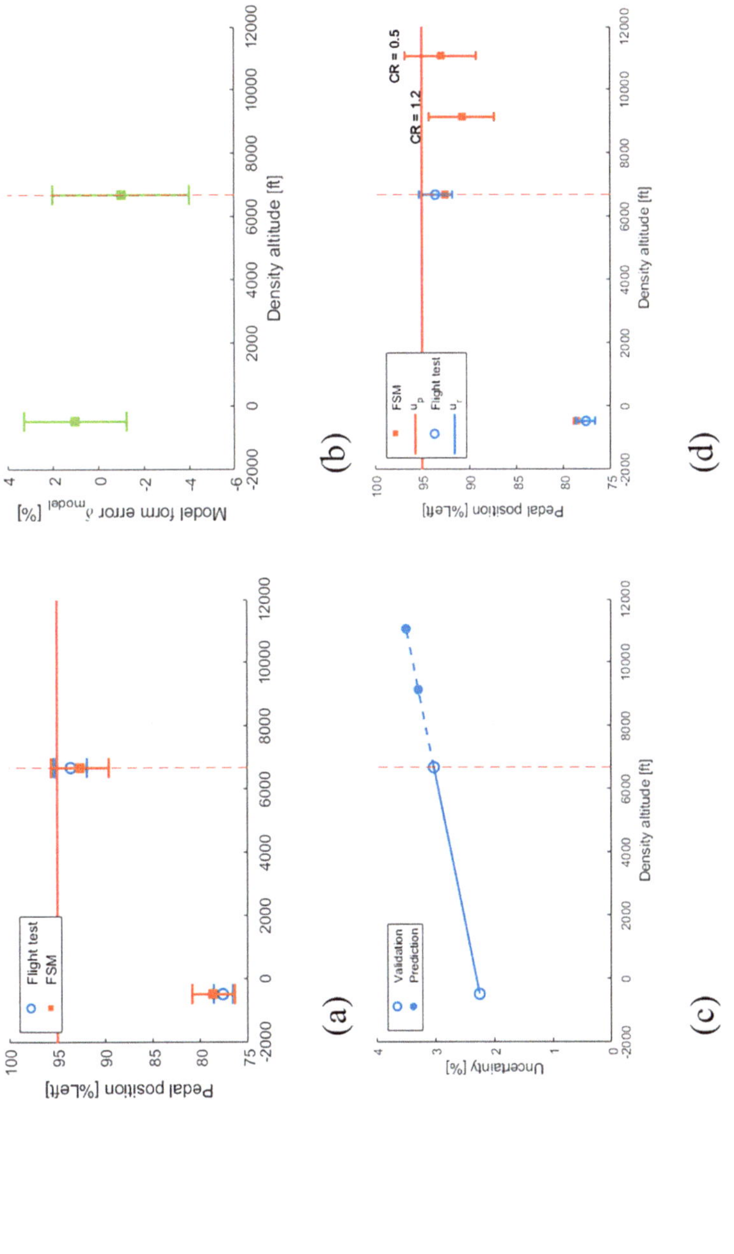

Fig. 9.4 Example of CR analysis; pedal margin for critical azimuth test in the DoE

In this example, the simulation prediction clearly does not meet the minimum requirement of a CR greater than unity at the 11,000 ft extrapolation. At this point, the applicant is faced with the choice of accepting a take-off gross weight restriction, or improving the modelling and/or reducing the prediction uncertainty. The latter may be achieved by gathering additional flight test data to reduce the amount of extrapolation, updating the model form of the simulation, or by endeavouring to reduce the input uncertainties to which the prediction is most sensitive.

Note that simple linear extrapolation of the model form uncertainty does not inherently result in increased uncertainty, although the data trends in that direction in this example. More rigorous methodologies such as regression fitting provide confidence bounds that quantify the uncertainty of the extrapolation, tending towards greater uncertainty with increasing extrapolation distance, see also Chap. 13.

9.3 Exploring DoV Fidelity Assessments Extrapolated into the DoE

Test data define the boundaries of, and are scattered throughout the DoV, but uncertainties prevail. Two such examples involving epistemic uncertainty are,

(a) Using the metrics defined in Phase 1, the initial comparison between flight and simulation show insufficient fidelity at an ACR condition. A model update process is undertaken to reduce the mismatch to within tolerance; but there is no statistical information available on the input parameters exploited in the model updating.

(b) Interpolation is used to derive fidelity at condition Y based on validation at conditions X and Z, where fidelity has been deemed sufficient. Uncertainty surrounds whether the (potentially different) model updates applicable at conditions X and Z, or some combination of both, will also be applicable at condition Y.

The potential for gaining insight into fidelity (model form errors) uncertainties is high in the DoV due to the multiple validation points analysed. A technique that has proved useful for the analysis and characterisation of epistemic uncertainty in fidelity assessment is described as 'renovation'. First, stability and control derivatives (see Chap. 6) predicted for the FSM are compared with those estimated from flight, e.g. using system identification techniques. Based on the comparisons, a subset of derivatives is selected that time and frequency domain response metrics are shown to be particularly sensitive to; the sensitivity being quantified by user-defined cost functions [9]. Such derivatives quantify local stability and response behaviour of the nonlinear FSM, and the p-model estimated from flight test. Cost function sensitivity can highlight the importance of individual derivatives to stability and/or response and the difference between the flight estimates and FSM predictions can be used to explore physical sources of fidelity deficiencies. In some cases, 'delta derivatives' can then be formed (difference between estimate and prediction) to augment the nonlinear FSM, and thence to improve fidelity. In a further step, an examination of the component (e.g., main rotor, fuselage, tail rotor, see Figs. 6.2 and 6.3) breakdown of the predicted FSM stability and control derivatives can potentially provide the required insight into the physical sources of modelling errors. Through this process, judgements can be made on the sources of modelling deficiencies. In principle, if the renovation can be shown to be suitable also at extrapolated conditions within the DoE, it can be used to minimise simulation error as well. This process is exercised in the Case Study reported in Chap. 14, and here an example is presented to assess the effectiveness of a renovation made at one flight condition, applied to another.

Example scenario, CS-27 and CS-29 requirements for dynamic stability. For rotorcraft to be approved for VFR, CS 27 has no specific requirements for dynamic stability while CS 29 only imposes very simple dynamic stability requirements for Category A approved rotorcraft (i.e. every short period oscillation occurring above the best rate of climb airspeed (V_y) must be damped). The standards set for IFR flight (CS 27 and 29, Appendix B) are more demanding and specify stability requirements in terms of damping ratio as a function of the period of oscillation. In Fig. 9.5, the boundaries for the various certification standards are shown on the frequency-damping chart together with the ADS-33 boundaries for Level 1, 2 and 3 Handling Qualities, and with data points for the RoCS AW109 case study on Dynamic Stability (DS), expanded on in Chap. 14. For this example, the validation process at one (altitude-velocity) point (3000 ft, level flight, 120 kts) in the DoV, revealed that the FSM prediction for the Lateral Directional Oscillation (LDO) was just outside the CS-27/29 boundary (x, TD F-AW109), while estimates from flight test (FT) showed the LDO to be just inside the certification boundary (*, TD FT DB Esti). A simple renovation determined that an FSM-update involving a 10% increase in the yaw damping derivative, N_r, was sufficient to bring this fidelity metric (+, RF-AW109(3000ft)) into the adopted sufficiency range, defined in this example as a 10% (blue-dashed) 'box' around the mean flight test point. The figure also shows uncertainty boxes wrapped around the flight test and simulation test points, based on the varying computations of frequency and damping using different sections of the pedal doublet-induced yaw response test data.[1] In the case of the simulation data, this included data derived from different control input magnitudes.

Figure 9.5 also shows the FSM prediction (LF-AW109) for the 10,000 ft case, together with the renovation (RF-AW109) after application of the same 10% yaw damping improvement that was effective at the lower altitude. The applicant might argue that there were no significant differences in the flight characteristics (and hence FSM form) at the two altitudes to justify additional updates, but they would, of course, need to offer explanations for the damping deficiency.

The data points discussed above are for the bare airframe, or SAS-off, configuration. As shown in the figure, there is a significant stability margin for the CS29 VFR ACR, and CS27 VFR ACR, for which there is no quantified stability requirement.

[1] In this example, the FSM and FT frequency-damping points for the 3 kft case on the eigenchart were both derived from time-domain computations of the yaw rate response to pedal doublet inputs.

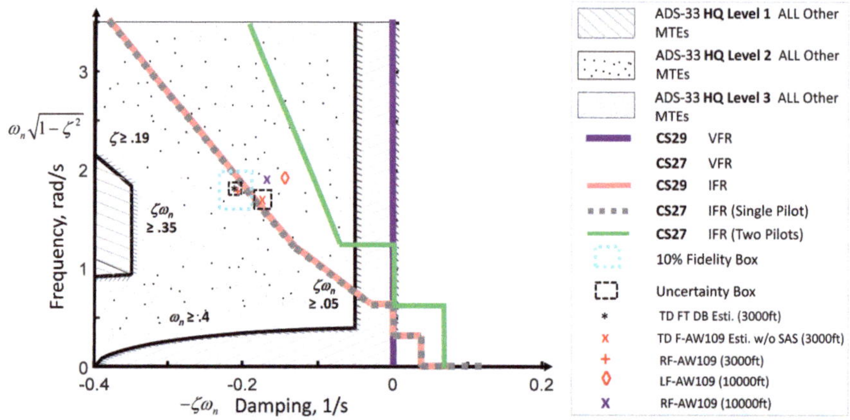

Fig. 9.5 FSM renovation to achieve sufficient fidelity for dynamic stability at 120 kts, with the corresponding DS boundaries; 3kft result compared with flight test, and 10 kft result with extrapolated renovation (TD—test doublet, FT—flight test, RF—renovated F-AW109, LF—linearised F-AW109, SAS—stability augmentation system)

However, uncertainties for the 3000ft case would make IFR certification questionable, with such a small stability 'margin'. The data suggest that, at the higher altitude, the aircraft would fail the IFR certification. This is, of course, not untypical of helicopters without stability augmentation.

In the above case, the renovations, or updates, were made using a single 'delta' derivative, augmenting the nonlinear FSM yaw damping with a 10% increase in N_r. A plausible physical explanation for this is that the wind tunnel tests to derive fuselage and empennage forces did not capture the interference/blockage effects correctly, both statically and dynamically. In addition, there are uncertainties regarding the modelling of the blockage effects on the tail rotor in the FSM. UQ could include varying interference modelling parameters, within the DoR range, to explore sensitivities, coupled with additional CFD analysis to compare with the wind tunnel test data.

While the dynamic stability example above has focused on a single deficiency source, within the rotorcraft FSM there are a multitude of sources of potential mismatch that have featured in the limited public-domain references on simulation fidelity. A good understanding of such potential model-form errors and their sources is considered critical to developing a high level of confidence in the RCbS process. Table 6.1 listed several potential sources of model-form errors, describing these as the 'usual suspects'.

9.4 Phase 3 Concluding Remarks

This Book recommends that Credibility Assessment should be contained within the final Certification activity of the RCbS process. As such, the way that results are presented to certification authorities needs to be documented in the PMP as part of the Certification Plan (Chap. 3). As with other elements of the RCbS process, this Book emphasises early-adopter experience to be shared community-wide to maximise the capturing of lessons-learned as RCbS practices evolve.

9.5 Summary: Phase 3

To summarise, activities in Phase 3, Credibility assessment and certification activities, include:

i. RCbS certification tests performed for relevant ACRs
ii. uncertainty characterisation and quantification undertaken throughout the domain of prediction,
iii. credibility analysis and assessments undertaken on the results of i. and ii.,
iv. results assembled and presented to certification authorities to make case for certification,
v. based on the feedback from Phases 2 and 3, the Requirements Specification is updated to constitute a formal element of the case for RCbS for the selected ACRs.
vi. Outputs; updated RCbS Requirements Specification and Type/Supplemental-Type Certificate documentation.

References

1. Anon (2016) Standard for models and simulations NASA-STD-7009A. NASA
2. ASME (2009) Standard for verification and validation in computational fluid dynamics and heat transfer. The American Society of Mechanical Engineers, New York, NY
3. Mehta UB, Eklund DR, Romero VJ, Pearce JA (2016) Simulation credibility: advances in verification, validation, and uncertainty quantification. NASA/TP—2016–219422
4. Anon (2021) Recommended practice: when flight modelling is used to reduce flight testing supporting aircraft certification. AIAA R-154-2021, AIAA
5. Mauery T (2021) A guide for aircraft certification by analysis. NASA/CR-20210015404
6. Anon (2014) AC 29-2C Certification of transport category rotorcraft. Federal Aviation Administration
7. van't Hoff S, White MD, Dadswell CM, Lu L, Padfield GD, Quaranta G, Podzus P (2023) Case studies to illustrate the rotorcraft certification by simulation process; CS 29/27 low-speed controllability. In: 49th European rotorcraft forum, Bückeberg, Germany, 5–7 September
8. Lu L, Padfield GD, White M, Quaranta G, van't Hoff S, Podzus P (2023) Case studies to illustrate the rotorcraft certification by simulation process; CS 29/27 dynamic stability requirements. In: 49th European rotorcraft forum, Bückeberg, Germany, 5–7 September

9. Lu L, Padfield GD, White M, Perfect P (2011) Fidelity enhancement of a rotorcraft simulation model through system identification. Aeronautical J 115(1170):453–470. https://doi.org/10.1017/S0001924000006102

Chapter 10
Resourcing the RCbS Process

It is recognised that building a capability able to fully embrace virtual engineering in certification, as summarised in Fig. 2.1, and expanded on throughout this Book, will take time and dedicated resources. The emphasis here on 'dedicated' is part of a recommendation, to ensure that capabilities are 'grown' without the constraints and pressures of current programmes.

Technical capabilities will include the disciplines of flight dynamics and control and associated multi-body dynamic modelling, aerodynamics/structural dynamics, aeroelastics, numerical modelling, software and systems engineering, flight simulation and associated flight simulator technologies, system identification methods and applications, flight and wind tunnel testing and measurement techniques etc.

Technical capabilities will feature large, and are of course critical, but other skills and experience will also be important and likely critical. Individuals whose strength is in completing and finishing a task, deciding when enough is enough, writing the PMP or certification case, or designing/conducting the flight trials, are unlikely to be the same people who develop the CFD codes or UQ algorithms, or try to make sense of the results of validation or credibility, whose understanding of the physics within rotorcraft aeromechanics needs to be at the highest level. The use of full authority active control, perhaps with autonomous capability and multiple failure modes, requires yet another expertise and skills-set.

The list of critical skills and experience is extensive and, as such, the RCbS team will have strength in breadth and depth, and, to be fully effective, will need to operate as a team, across multiple disciplines and departments, with heightened awareness of important synergies and the need to practice within a systems-engineering discipline. The team will also need to practice 'as if' the FSM and FS were the 'real aircraft', so experience from the 'real world' will be essential. How such a team is developed and operates within the extant structure of a rotorcraft manufacturer is an important question. This Book has not offered specific guidance on this. Many of the RCbS tools and capabilities will be beneficial at the design and development stages of the rotorcraft life-cycle, and some will already be in place. However, there is also benefit

G.D. Padfield et al., *Rotorcraft Certification by Simulation and Analysis*, Springer Aerospace Technology, https://doi.org/10.1007/978-3-031-86398-1_10

in the team having a degree of autonomy from, say, the Chief Designer's team or the Airworthiness or Flight Test Departments. There are likely to be synergistic project timelines and milestones, so close collaboration is essential. But, with a focus on ACRs and the Certification Plan, the RCbS team needs to be able to effectively act as consultants to the design teams across a OEM's product range.

Major challenges are to establish what is required to achieve 'sufficient' fidelity along with maximum acceptable uncertainty, and to quantify credibility in the Company's flight simulation models, flight simulators and flight test measurement systems, to make a significant impact on certification costs, timescales and safety. Developing a profound understanding of what is meant by sufficient and credible, is part of these challenges.

Such challenges can be approached by first applying the RCbS process at various I-P levels, and for specific ACRs, to existing certified products, taking advantage of existing flight test databases. This would also provide opportunities to train new engineers in the exercising of the RCbS phases. In this way, the existing 'operational' capabilities can be drawn on, providing a framework for the RCbS process development. Also, such extant capabilities can be improved as new methods, focused on the certification application, are developed. As the process matures to the point where it can be applied to a new application, a question that might arise is, 'how deep should the RCbS capability be', particularly in terms of dissimilar redundancy? Because of the importance of the V&V processes in Phase 2, a strong argument could be made for at least 'duplex' in each discipline. This would enable a progressive and independent checking of any analysis and the results of simulation. This could also extend to the simulation tools adopted in the process, e.g. CFD codes, flight models, system identification techniques. The dedicated RCbS team will also face a significant challenge when using 'existing' flight test data, usually captured for unrelated purposes. It is suggested that dedicated test programs in support of RCbS capability development will also be required to ensure that validation and fidelity assessment processes can be productively exercised and refined.

Early applications are likely to be modest in their aims, but it is strongly recommended that the long-term aspiration to achieve the I-P full credit goals define the backbone of the capability development. These aspects are elaborated on in Chaps. 12–14 of the Book, and revisited in the wrap-up Chap. 15.

Chapter 11
Guidance for Selected ACRs Within the Certification Specifications

11.1 Introduction

The RCbS process documented in this Book has been partially 'exercised' using the University of Liverpool's HELIFLIGHT-R simulation facility (Fig. 11.1), on three CS27/29 ACRs and reported as Case Studies in Chaps. 12, 13, and 14. In a 'perfect world', the exercises would follow the advocated RCbS process closely, through assembling the PMP within the Certification Plan in Phase 0, to Credibility assessment and (Virtual) Certification in Phase 3, highlighting lessons learned. In this 'perfect-world' scenario, physics-based updating strategies to achieve FSM and FS fidelity sufficiency in the DoV would be presented, along with uncertainty quantification and confidence analysis for results in the DoE. Following the RCbS process requires that the 'requirements' for the FSM, FS and FTMS are developed in concert in Phase 1, for the ACRs and I-P levels being considered, followed by the build and development activities for all three elements in Phase 2. The so-called pre-certification flight testing would take place in parallel with the FSM/FS/FTMS build, V&V and fidelity assessment in Phase 2. We describe the exercising as 'partial', because such extensive work was not possible within the timescales and resources of the RoCS project, and hence achievement objectives had to be bounded.

The FLIGHTLAB FSM and flight test data for the AW109 Trekker were provided to the RoCS team with comparative assessments for trim (controls and attitudes), stability (eigenvalues) and gain/phase frequency responses, across a range of straight and level flight conditions and aircraft weight and balance configurations. Fidelity assessments to establish sufficiency for the selected ACRs were carried out as part of the Case Studies. In the early stages of the RoCS project, an assessment was made of the ACRs within the relevant sections of CS 27 and CS 29 to agree a short list of candidates for the exercising. Several factors came into play here including the availability of flight test data, the anticipated FSM and FS capabilities, expertise and experience within the RoCS team and, of course, the estimated resource requirements. Four different flight simulators were used during the RoCS project, at DLR, NLR,

G.D. Padfield et al., *Rotorcraft Certification by Simulation and Analysis*, Springer Aerospace Technology, https://doi.org/10.1007/978-3-031-86398-1_11

Fig. 11.1 HELIFIGHT-R (foreground); views from crew station (Right)

LHD and UoL, and results from early trials at DLR and NLR reported (e.g. [1]). Liverpool's HELIFLIGHT-R FS was selected for the penultimate exercising of the RCbS process in July 2023, when the three selected Case Studies were investigated. Section 11.2 of this Chapter provides a description of the HELIFLIGHT-R facility and the various rating scales and the in-cockpit-questionnaire used during the trial.

The I-P conditions chosen for using simulation as a MOC for the selected ACRs were deliberately 'ambitious'. The RoCS team wanted to demonstrate success in some cases, but also failure in others, to bring out aspects that would need strengthening in a 'real' application.

Within the trim-stability-response fidelity framework described in Chap. 4, and in broad terms, Case Study 1 focuses on Trim, Case Study 2 on Response and Case Study 3 on Stability. Details of the exercises are reported in Chaps. 12, 13 and 14. Below, some key observations, conclusions and recommendations are summarised for each of the Case Studies.

Case 1: Low-Speed Controllability and Manoeuvrability. This Case Study explores low speed handling qualities with the applicable CS-29 sections §29.143(c-d) requiring that, "*wind velocities from zero to at least 31 km/h (17 knots), from all azimuths, must be established in which the rotorcraft can be operated without loss of control on or near the ground in any manoeuvre appropriate to the type (such as crosswind take-offs, sideward flight, and rearward flight).*" The conventional test technique flown to demonstrate compliance involves the aircraft following a pace-car (or pace-car equivalent (PCE) using cockpit displays) on a runway to establish the required sideslip conditions. The pilot then applies pedal inputs to demonstrate controllability and manoeuvrability (C&M). The expectations are that at least two 'critical' wind azimuths will emerge; one related to pedal margin, typically with winds from around 90° starboard (for anti-clockwise rotors), the other related to handling deficiencies, with winds from somewhere in the range 225–270°. The Case Study focussed on the first of these since it was suspected that the modelling of tail rotor vortex-ring-state would not provide sufficient yaw control fidelity to address the second. For this I4-P3 scenario, the opportunity was taken to explore several sources of pedal margin prediction uncertainty for extrapolation to high density altitude; as described in Chap. 12, prediction uncertainty unavoidably comes at the expense of performance margin. A new flight test manoeuvre was designed, considered more representative of the operational scenario for which low speed C&M needs to be demonstrated. The cross-wind-hover (XWH) required the pilot to establish a hover with defined wind strength and direction and to exercise yaw control in a similar manner to the PCE. A comparison of results from the two FTMs showed differences in control activity/compensation and achieved task performance for the same wind conditions; the XWH being the more difficult. While no definite conclusions can be drawn from this comparison, the Case Study illustrates how flight simulation provides the opportunity to explore such variations in the demonstration of compliance with an ACR.

Case 2: Category A Take-off and Landing. This is an ACR featuring in several CS-29 sections: §29.49, §29.53–62, §29.67, §29.77–81, §29.85, §29.141(b), §29.143(e) addressing CAT A Rejected takeoff in a confined area:

The rejected take-off distance and procedures for each condition where take-off is approved will be established with the take-off path requirements of CS 29.59 and CS 29.60 being used up to the TDP (Take-off Decision Point) *where the critical engine failure is recognised, and the rotorcraft landed and brought to a stop on the take-off surface etc.*

In this Case Study, described in more detail in Chap. 13, the applicant aims to use simulation to determine the maximum take-off weight for the ACR, corresponding to I4 (Full Credit), and to extrapolate from flight test data obtained in the low altitude (748 ft) Domain of Validation (DoV) to a high-altitude case (9280 ft), so predictability level P3, "*Extensive extrapolation in the Domain of Extrapolation (DoE)*".

The FTM designed for testing compliance with this ACR involved the take-off, rearward climb to the decision point where one engine fails, the pilot then stops the climb and establishes a descent to cushion and touchdown on the helipad. Clearly, several phases are involved in this FTM in which the pilot was required to assess the different handling qualities that informed compliance judgements. Prior to the testing at Liverpool, fidelity assessments were conducted using a virtual pilot model to determine the required sufficiency for realistic control strategies. During the HELIFLIGHT-R trial, vestibular motion cueing was found to be important during the final cushion and touchdown phase. One outcome of the testing was that the EP considered that "*full credit*" could be expected for this ACR in the DoE at a mass of

2505 kg at the extrapolated high-altitude condition. An issue that stimulated discussion, that is sure to continue, was how the HQ assessment, and awarded HQR, might be used in the compliance demonstration, following on from the fidelity assessment? What would be an acceptable HQR in the context of the 'means of compliance'? The RoCS project did not offer a definitive answer to this important question.

Case 3: VFR/IFR Dynamic Stability. Within CS 27/29, the relevant ACR is CS 29.181 and, also, for IFR-flight, CS 27/29 Appendix B. For the most part, dynamic stability is quantified in terms of relative damping as a function of the frequency of an aircraft's oscillatory modes of motion. Using simulation to demonstrate compliance is relatively straightforward with the 'off-line' application of clinical control inputs used to excite oscillations. The opportunity was taken to assess extrapolation from the DoV reference condition (120 kts straight and level at 3000 ft) to higher altitude, and variations in airspeed and climb rate. Stability derivatives were used to inform both the model update process in the DoV and the assessment of credibility in the DoE. The CSs do not require closed-loop testing for demonstrating compliance with the dynamic stability ACR. To explore whether such testing could be of value, a 45° turn FTM was designed to explore pilot subjective assessments of the handling qualities, in the presence of turbulence, to compare with the results from off-line testing. This highlighted the value of piloted simulation, in being able to set up well-defined, repeatable, conditions for pilot assessment of compliance.

The Case Study is reported in Chap. 14, highlighting a potential for significant replacement of flight testing with simulation across a wide range of P3 and P4 levels.

11.2 The HELIFLIGHT-R Flight Simulation Facility and Rating Scales Used in the RoCS Piloted Simulation Trials

In 2008, the HELIFLIGHT-R facility [2] (Fig. 11.1) was commissioned at the University of Liverpool, to complement the smaller, single seat, HELIFLIGHT simulator [3], and became the workhorse for flight dynamics, control, handling qualities and simulation research. With its much larger motion envelope (Table 11.1) and wider/deeper visual field-of-view (Fig. 11.2), the facility provided higher fidelity for applications where these forms of motion cueing are vitally important, e.g. at the helicopter-ship dynamic interface, close to surface operations. Reference [2] describes the processes adopted during the development and acceptance testing of HELIFLIGHT-R. References [4–15] detail some of the results of research using HELIFIGHT-R in its first decade of operation related to flight simulation fidelity and motion cueing research, together with its use in applied research for aircraft clearances and certification activities. Reference [11] provides details of the Simulation Fidelity Rating (SFR) scale that was developed to assess the utility of a Flight Simulator (FS) for training purposes. A modification to the SFR scale for certification applications is shown in Fig. 11.3 with the text highlighted in red indicating suggested changes; further work is needed to confirm the veracity of the proposed changes.

HELIFLIGHT-R underwent significant upgrades in 2016 to increase projector resolution, and again in 2021 to add a fourth projector to increase downward field of view, HELIFLIGHT-R+ (Fig. 11.1). The instrument panels were also updated in 2021 to large, higher-resolution, multi-touch displays to enable use of modern glass-cockpit interfaces.

The various Rating Scales and In-Cockpit-Questionnaire adopted during the Liverpool trials are shown in Figs. 11.4, 11.5, 11.6 and 11.7.

Simulation Fidelity Rating Scale (Ref. [11])

Simulation Fidelity Rating Scale Terminology and Process for Awarding Ratings

Table 11.1 HELIFLIGHT-R motion capability

Property	Displacement	Velocity	Acceleration
Pitch	−23.3°/25.6°	± 34°/s	300°/s^2
Roll	−23.2°	± 35°/s	300°/s^2
Yaw	± 24.3°	± 36°/s	500°/s^2
Heave	± 0.39 m	± 0.7 m/s	± 1.02 g
Surge	−0.46/ + 0.57 m	± 0.7 m/s	± 0.71 g
Sway	± 0.47 m	± 0.5 m/s	± 0.71 g

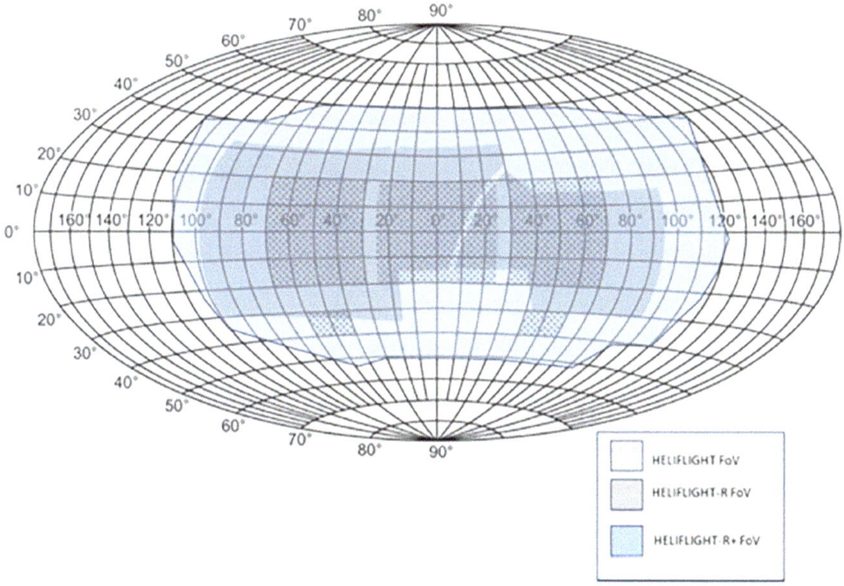

Fig. 11.2 HELIFLIGHT-R, HELIFLIGHT-R+ and HELIFLIGHT field-of-view

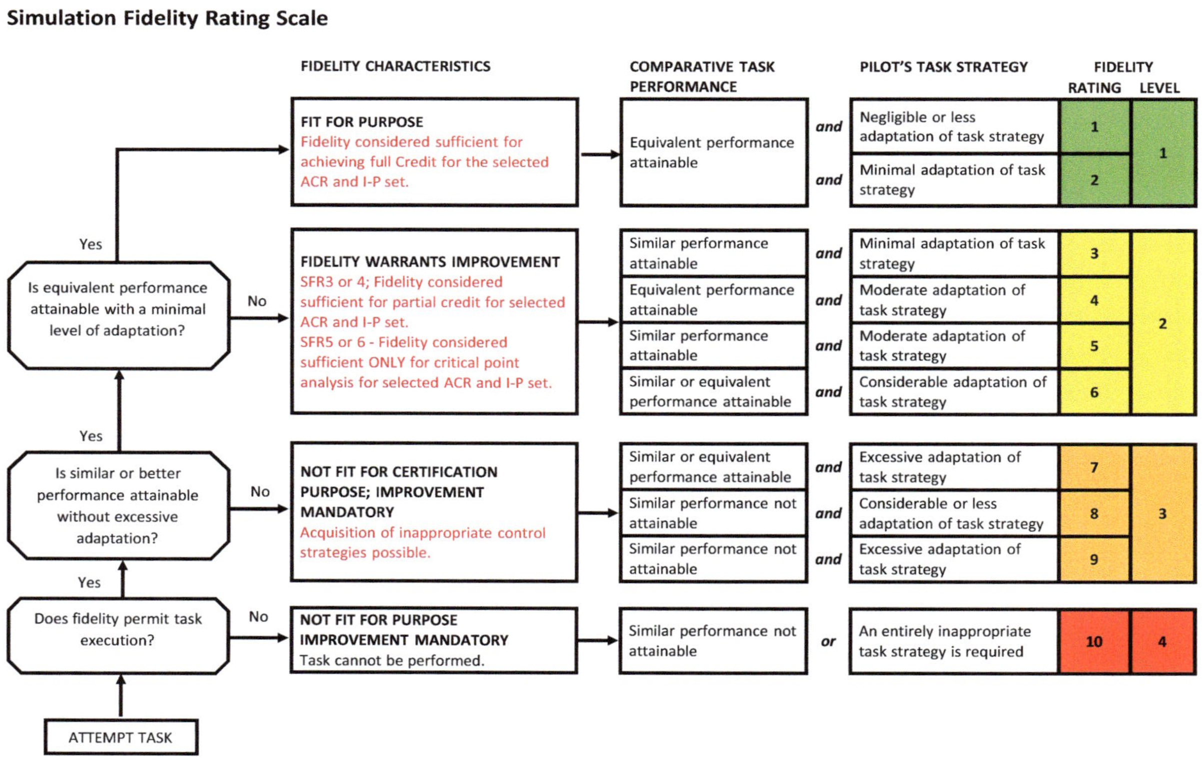

Fig. 11.3 Simulation fidelity rating scale [text in red used for application to certification]

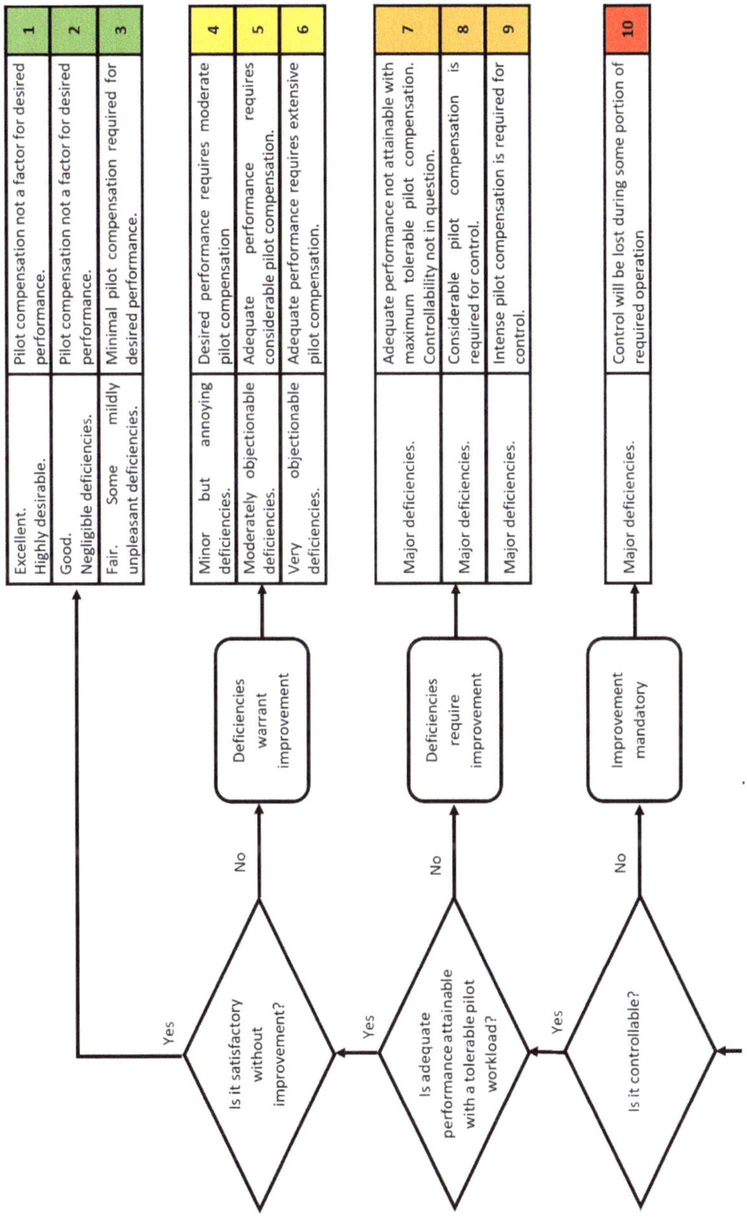

Fig. 11.4 Cooper-Harper handling qualities rating scale

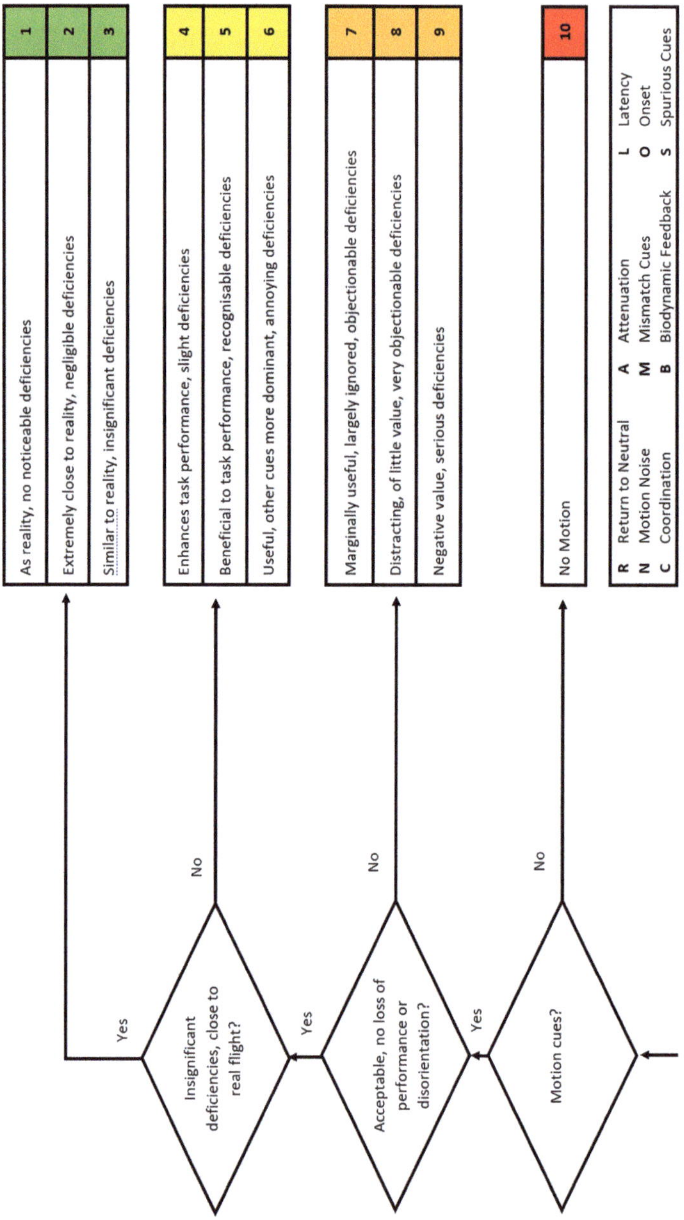

Fig. 11.5 Motion fidelity rating scale

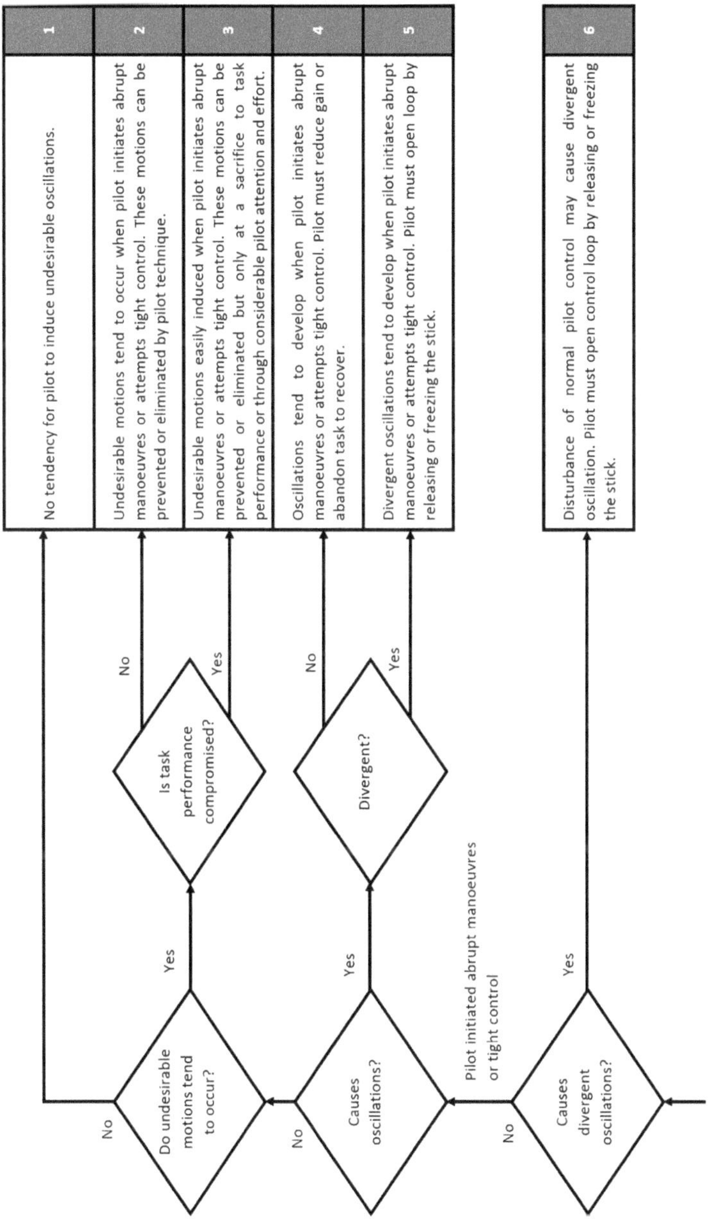

Fig. 11.6 Pilot opinion scale for pilot induced oscillations

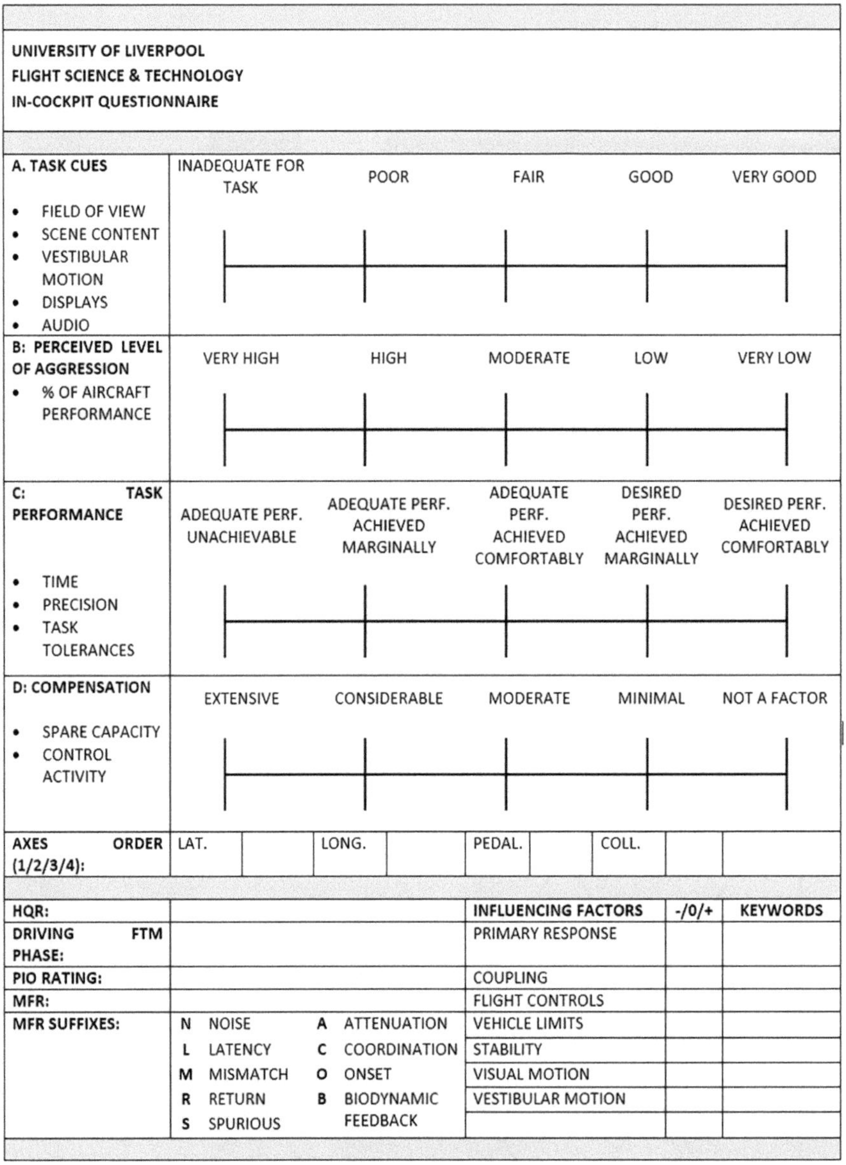

UNIVERSITY OF LIVERPOOL
FLIGHT SCIENCE & TECHNOLOGY
IN-COCKPIT QUESTIONNAIRE

A. TASK CUES • FIELD OF VIEW • SCENE CONTENT • VESTIBULAR MOTION • DISPLAYS • AUDIO	INADEQUATE FOR TASK	POOR	FAIR	GOOD	VERY GOOD
B: PERCEIVED LEVEL OF AGGRESSION • % OF AIRCRAFT PERFORMANCE	VERY HIGH	HIGH	MODERATE	LOW	VERY LOW
C: TASK PERFORMANCE • TIME • PRECISION • TASK TOLERANCES	ADEQUATE PERF. UNACHIEVABLE	ADEQUATE PERF. ACHIEVED MARGINALLY	ADEQUATE PERF. ACHIEVED COMFORTABLY	DESIRED PERF. ACHIEVED MARGINALLY	DESIRED PERF. ACHIEVED COMFORTABLY
D: COMPENSATION • SPARE CAPACITY • CONTROL ACTIVITY	EXTENSIVE	CONSIDERABLE	MODERATE	MINIMAL	NOT A FACTOR

AXES ORDER (1/2/3/4):	LAT.		LONG.		PEDAL.		COLL.		

HQR:		INFLUENCING FACTORS	-/0/+	KEYWORDS
DRIVING FTM PHASE:		PRIMARY RESPONSE		
PIO RATING:		COUPLING		
MFR:		FLIGHT CONTROLS		
MFR SUFFIXES:	N NOISE L LATENCY M MISMATCH R RETURN S SPURIOUS	A ATTENUATION C COORDINATION O ONSET B BIODYNAMIC FEEDBACK	VEHICLE LIMITS STABILITY VISUAL MOTION VESTIBULAR MOTION	

Fig. 11.7 University of Liverpool in-cockpit questionnaire

Performance:

- **Equivalent Performance**: The same level of task performance (desired, adequate etc.) is achieved for all defined parameters in simulator and flight. Any variations in performance are small.

- **Similar Performance**: There are no large single variations in task performance, or, there are no combinations of multiple moderate variations across the defined parameters.
- **Not Similar Performance**: Any large single variation in task performance, or multiple moderate variations, will put the comparison of performance into this category.

Adaptation:

- **Control Strategy**: changes in the level of compensation, including in the size, shape and frequency of the applied control inputs
- **Cueing**: differences in the way in which task cues are presented to the pilot
- **Workload**: including differences in the physical effort of moving the controls; scanning of the available task cues; and the mental work associated with interpreting cues and determining control inputs.
- **Vehicle Response**: differences in the perceived response of the vehicle

Process for awarding the SFR and HQR in the assessment of simulation fidelity

(1) Test team ensures that the pilot is briefed on the FTM and is familiar with the in-cockpit questionnaire (Fig. 11.7) and the ratings scales (Figs.11.3, 11.4, 11.5 and 11.6) to be used
(2) Emphasise that the SFR is a comparative rating; simulation compared with flight
(3) Pilot flies an FTM in the aircraft, enough times that s/he is clear what the HQ deficiencies are and is ready to complete the questionnaire
(4) Pilot completes the questionnaire; key entries are C (task performance) and D (level of compensation needed), driving FTM phase and influencing factors
(5) Pilot awards the HQR (Fig. 11.4), consistent with the (performance and compensation) entries in the questionnaire, clearly identifying the driving phase and HQ problems. If other rating scales are also being used (MFR (11.5), PIO (11.6)) the ratings are also awarded and recorded on the questionnaire.
(6) Pilot flies the same FTM in the flight simulator, enough times that s/he is clear what the HQ AND SF deficiencies are and is ready to complete the questionnaire
(7) Pilot completes the questionnaire as for flight case
(8) Pilot awards the HQR and SFR (Fig. 11.3), consistent with the entries in the questionnaire
(9) While the pilot is not expected to be able to quantify the HQ or SF deficiencies in technical terms, the dialogue between pilot and engineer around the 'influencing factors' in the questionnaire is intended to inform the identification of what and how, e.g. improvements are warranted

Cooper-Harper Handling Qualities Rating Scale (Ref. [16])

Motion Fidelity Rating Scale (Ref. [13])

Pilot Opinion Scale for Pilot Induced Oscillations (Ref. [17])

References

1. Podzus P, White MD, Dadswell CM, van't Hoff S, Crijnen J (2022) Evaluation of simulator cueing fidelity for rotorcraft certification by simulation. In: 78th VFS Annual Forum & Technology Display, Fort Worth, Texas, May 10–12
2. White MD, Perfect P, Padfield GD, Gubbels AW, Berryman AC (2013) Acceptance testing and commissioning of a flight simulator for rotorcraft simulation fidelity research. Proc Inst Mech Eng Part G: J Aerospace Eng 227(4):663–686. https://doi.org/10.1177/0954410012439816
3. Padfield GD, White MD (2003) Flight simulation in academia: HELIFLIGHT in its first year of operation. Aeronautical J 107(1075):529–538. https://doi.org/10.1017/S0001924000013415
4. Lu L, Padfield GD, Jump M (2010) Investigation of rotorcraft-pilot couplings under flight-path constraint below the minimum-power speed. Aeronautical J 114(1157):405–416. https://doi.org/10.1017/S0001924000003882
5. Lu L, Padfield GD, Perfect P, White MD (2011) Fidelity enhancement of rotorcraft simulation through system identification. Aeronautical J 115(1170):453–470. https://doi.org/10.1017/S0001924000006102
6. Padfield GD (2011) The Tau of flight control. Aeronautical J 115(1171):521–556. https://doi.org/10.1017/S0001924000006187
7. Padfield GD, Lu L, Jump M (2012) Optical Tau in boundary avoidance tracking—a new perspective on pilot induced oscillations. AIAA J Guidance Control Dyn 35(1):80–92. https://doi.org/10.2514/1.54065
8. Forrest J, Owen I, Padfield GD (2012) Ship/Helicopter Operating Limit (SHOL) prediction using piloted flight simulation and time-accurate airwakes. AIAA J Aircraft 49(4):1020–1031. https://doi.org/10.2514/1.C031525
9. Hodge SJ, Forrest JS, Padfield GD, Owen I, Spencer MP (2012) Simulating the environment at the aircraft-ship dynamic interface: development & application. Aeronautical J 116(1185):1155–1184. https://doi.org/10.1017/S0001924000007545
10. Perfect P, White MD, Padfield GD, Gubbels AW (2013) Rotorcraft simulation fidelity: new methods for quantification and assessment. Aeronautical J 117(1189):235–282. https://doi.org/10.1017/S0001924000007983
11. Perfect P, Timson E, White MD, Padfield GD, Erdos R, Gubbels AW (2014) A fidelity scale for the subjective assessment of simulation fidelity. Aeronautical J 118(1206):953–974. https://doi.org/10.1017/S0001924000009635
12. Hodge SJ, Perfect P, Padfield GD, White MD (2015) Optimising the roll-sway motion cues available from a short stroke hexapod motion platform. Aeronautical J 119(1211):23–44. https://doi.org/10.1017/S000192400001023X
13. Hodge SJ, Perfect P, Padfield GD, White MD (2015) Optimising the yaw motion cues available from a short stroke hexapod motion platform. Aeronautical J 119(1211):1–21. https://doi.org/10.1017/S0001924000010228
14. Owen I, White MD, Padfield GD, Hodge SJ (2017) A virtual engineering approach to the helicopter-ship dynamic interface—a decade of modelling and simulation research at the University of Liverpool. Aeronautical J 121(1246):1577–1593. https://doi.org/10.1017/aer.2017.102
15. Memon WA, Owen I, White MD (2020) SIMSHOL: a predictive simulation tool to inform ship-helicopter clearance trials. AIAA J Aircraft 57(5):854–875. https://doi.org/10.2514/1.C035677
16. Cooper GE, Harper RP (1969) The use of pilot ratings in the evaluation of aircraft handling qualities, NASA TN D-5153. NASA
17. Anon (1997) Flying qualities of piloted aircraft, MIL-HDBK 1797. US Department of Défense Handbook

Chapter 12
Case Study 1: CS-27/29 Low Speed Controllability and Manoeuvrability

12.1 Introduction

This chapter presents the results from the Case Study on the Applicable Certification Rule (ACR) Certification Specification (CS) 27/29.143(c), 'controllability and manoeuvrability in winds up to 17 knots' [1, 2] to illustrate the application of the guidance presented in this Book. To focus attention, CS-27.143(c) states that:

> Wind velocities from zero to at least 31 km/h (17 knots), from all azimuths, must be established in which the rotorcraft can be operated without loss of control on or near the ground in any manoeuvre appropriate to the type (such as crosswind take-offs, sideward flight, and rearward flight), with:
>
> (1) Critical weight;
> (2) Critical centre of gravity;
> (3) Critical rotor rpm; and
> (4) Altitude, from standard sea-level conditions to the maximum take-off and landing altitude capability of the rotorcraft.

Although the requirements relate to 'any manoeuvre appropriate to the type' (including take-offs and landings), the focus in this Case Study is on hover in cross-wind conditions in ground effect (IGE). It must be demonstrated that an IGE hover in wind from any direction can be established, maintained and transitioned to/from without loss of control. The guidance in the Federal Aviation Administration's (FAA's) Advisory Circular (AC) 29-2C [3] further states that it must be demonstrated that sufficient control should be available to permit a clearly recognisable yaw response with wind from the critical azimuth(s). The applicant must then define the weight, altitude and temperature limits within which the requirements are met. For a CS 29 certified rotorcraft, CS 29.143 (c) must be demonstrated up to the maximum take-off and landing capability of the rotorcraft. CS 27.143 (c) is similar in content, with the only difference being that the demonstration must be carried out up to 7000 ft density altitude, or the maximum take-off and landing capability (whichever is less).

G.D. Padfield et al., *Rotorcraft Certification by Simulation and Analysis*, Springer Aerospace Technology, https://doi.org/10.1007/978-3-031-86398-1_12

Flight testing for compliance demonstration for the low-speed controllability and manoeuvrability (C&M) requirements is traditionally performed in calm winds using a pace car (or equivalent) over a runway for visual reference. Although it is questionable as to whether the conditions in this testing approach are truly equivalent to operational cross-wind hover tasks, this method is often preferred over real wind hover testing because of the difficulties in obtaining the desired steady wind conditions. In either case, the demonstration of controllability and manoeuvrability, or the ability to affect a recognisable yaw response, typically involves demonstrating transient heading changes in the order of $\pm$ 5° from the critical direction and back. The term 'critical conditions' is used to include any combination of wind magnitude and direction where control margins or handling qualities effects limit C&M.

In traditional development and certification, the flight test effort required to characterise the C&M of the aircraft, and ultimately demonstrate compliance with the certification requirements, is extensive. In particular, the requirement to demonstrate compliance up to the maximum take-off and landing altitude capability (or up to 7000 ft density altitude) implies a time-consuming and costly relocation to high-altitude test sites. Although the high-altitude test activities may be combined with various other test objectives, and cannot be avoided entirely, reduction in the time spent on-site and minimisation of the likelihood of encountering unexpected design limitations are considered highly desirable. Moreover, extensive simulation efforts in advance of flight testing, either off-line or in a flight simulator, reduce the risks associated with any remaining flight testing. Despite the low-speed nature, the IGE testing for this requirement is performed partly inside or close to the limits of the height-velocity diagram and is, therefore, of inherently high risk. Hence, the risk of loss of control at the pedal margin limits or from yaw handling deficiencies is another motivation for the use of simulation to mitigate the safety risks in the context of this ACR.

With these objectives in mind, what remains is to develop and demonstrate the ability to predict the behaviour of the aircraft in the relevant flight conditions with sufficient fidelity and credibility at a cost, in time and effort, that is not prohibitive. The following sections explore the various Phases of the RCbS process in more detail, including results from exploratory piloted simulation trials. Finally, preliminary conclusions and recommendations for ongoing work are presented based on the results of this Case Study.

12.2 Phase 1: Requirements Capture and Build

Influence and Predictability. In RCbS Phase 1, the applicant and authority must agree on the extent to which simulation will be relied upon in the finding of compliance with the ACRs. This includes an expectation as to how far the predictions may be extended beyond the validation flight test data in whatever extrapolation dimension may be applicable. It is assumed for this Case Study that the applicant desires to use simulation to determine the critical wind speeds and directions, and associated take-off weight limitations, accounting for the C&M requirements for IGE hover. Flight testing for low-speed C&M at the maximum altitude capability is not intended to be performed. Thus, the Influence level corresponds to I4, P3 (extensive extrapolation, full credit (see Tables 4.2, 4.3 and Fig. 4.4). It is recognised that this is a challenging scenario, one of the ambitious cases addressed to exercise the process, knowing that there would be degrees of both success and failure. The simulations are planned for conditions up to the envisioned maximum take-off and landing altitude at 12,000 ft density altitude. Initial flight testing for demonstration and simulation validation is planned to be performed up to 7000 ft (considering local geography). AC 29-2C [3] allows altitude extrapolation of maximum ± 2000 ft density altitude for IGE controllability assessment, putting this use case clearly in the P3, "Extensive extrapolation in DoE", predictability level. However, if limited high-altitude flight testing is later performed, e.g., to provide evidence for the absence of limitations due to exhaust gas re-ingestion and compressor surge, such testing would ultimately provide additional validation data. This data would then significantly reduce the extent of the extrapolation for an iteration on the validation and simulation predictions if needed. In any case, should the P3 level be considered too challenging, an applicant may choose to take advantage of being at a medium/high-altitude test site to gather validation data (and, thus, reduce the amount of extrapolation to a P2 level), while still relying on simulation to cover the full envelope.

Initial FSM and FS Requirements. Having established the conditions and objectives for the use of simulation as part of the Certification Plan, the next step is to establish, as far as can be foreseen in advance, the requirements that the Flight Simulation Model (FSM), the Flight Simulator (FS) and the validation flight test data must meet. The prediction, through simulation, of the available control power[1] at the critical azimuths, e.g., with wind from around 90°, may be performed off-line (although piloted simulation may serve to explore FSM anomalies). A controllability assessment aimed at limitations due to a deterioration in handling qualities, typically at azimuths characterised by significant vortex wake interactions, and/or tail rotor vortex ring state (VRS), requires piloted simulation in a suitable FS, although off-line predictions of stability and response can be used to guide the FS trials. The applicant, therefore, plans to employ both off-line desktop and piloted simulations (with the FS running either with the same or a derived and cross-correlated real-time FSM).

The following 'high-level' FSM characteristics are anticipated by the applicant to be required at a minimum to achieve acceptable credibility:

- Physics-based modelling (incl. extrapolation)
- Rigid blade element main and tail rotor
- Rotor wake interference effects on fuselage, empennage and tail rotor
- Dynamic VRS (tail rotor)
- Ground effect

[1] Control power is defined as the maximum response achievable at a trim condition, in terms of the appropriate response type, e.g., for a rate command response type, the control power is quantified in rad/sec.

- Engine deck: installed minimum spec. power
- Rotorspeed governor dynamics
- Uniform steady atmospheric wind.

It is assumed that the engine deck is provided by the engine OEM and takes into account corrections for installation effects. The requirements on the accuracy of the engine deck in terms of predicted available power at altitude in various wind directions will depend on the proximity to the engine operating limits. The significance of the rotorspeed governor dynamics in terms of their impact on controllability, particularly at high density altitudes, will need to be investigated for the specific aircraft through simulation and testing in the Domain of Validation (DoV).

One of the key difficulties for this particular application, at least for a conventional main-tail rotor configuration, is the modelling of the wake interactions between the rotors, fuselage, empennage and the ground. With contemporary computing power, real-time piloted simulations will typically require some form of reduced order modelling, e.g., in the form of data maps. VRS is mentioned here specifically because typical momentum theory modelling is aimed primarily at avoiding a singularity or discontinuity in the solution of the rotor inflow model. The modelling often does not capture the low-frequency unsteady nature, or the time lag associated with the development of the rotor inflow [4–6].

Prior to establishing preliminary requirements for the flight simulator, the applicant must define the way the piloted trials are to be performed. In the case of the low-speed C&M ACR, validation using Handling Qualities Ratings (HQRs) awarded in the simulator will require a scenario, or Flight Test Manoeuvre (FTM), that is comparable in terms of task and cueing environment to the flight test procedure. Thus, at least the simulator trials performed for validation purposes will be conducted in a zero-wind pace car type of set-up (see Appendix 1 for proposed FTM). However, as suggested above, the FS environment offers the opportunity to develop an FTM that is more comparable to the operational tasks associated with the ACR. Hence, for this Case Study the applicant proposes a 'cross-wind hover' (XWH) FTM for the predictions at altitude (see Appendix 2).

The following 'high-level' flight simulator characteristics are anticipated by the applicant to be required at a minimum to achieve sufficient (pilot) perceived simulation fidelity for this scenario [see simulation fidelity rating scale (SFR), Fig. 11.3]:

- Pilot inceptors with appropriate control loading
- Vestibular motion cueing: perhaps non-essential, but to be verified
- Visual cueing environment: sufficient to perform task with minimal pilot adaptation
- Vibration and audio cues: perhaps non-essential, but to be verified
- Generic instruments.

The visual cueing environment is influenced by many characteristics of the FS set-up, including the field of view, ground textures and contrast, and the visual content (i.e. objects). In the case of the proposed XWH FTM, artificial cues similar to those included in the precision hover Mission Task Element (MTE), defined in ADS-33E-PRF (ADS-33) [7], are required, where the definition of the adequate and desired performance standards needs to be related to typical civil operations. Although standardised to an extent (e.g. in terms of artificial visual cues), the sufficiency of the cueing environment for the specific ACR should be investigated early in the RCbS process through exploratory FS trials. The cockpit instruments do not need to be exact hardware or software replicas of the production aircraft, but it is considered essential that the scanning pattern and sequence match those observed in flight, which can be evaluated, e.g., through eye tracking measurements.

Vestibular motion cueing, although considered by the applicant to be non-essential in principle, can be useful in providing early cueing of, e.g., initial acceleration prior to lateral drift, thereby improving task performance and reducing pilot adaptation. The absence of such cueing may be considered conservative (i.e. flying the simulator is more difficult than real flight), but care must be taken to ensure that the pilot adaptation does not significantly influence the handling qualities evaluation. To avoid adverse vestibular motion cueing, it is essential that the motion system is properly tuned for the specific manoeuvre being investigated. Note that vestibular motion cueing includes vibration cues (even if the vibration cues may be provided by a separate system, such as a seat shaker). In some cases, the limiting factor in terms of aircraft controllability in the most critical conditions may be due to excessive vibrations. The propensity towards such characteristics, if identified during DoV flight testing, should guide the simulator requirements in terms of vestibular motion cueing.

Beyond establishing the objectives for the use of flight simulation and defining an initial set of simulation requirements, the applicant must put forth a plan at the end of Phase 1 that outlines the approach to, and data requirements for, Verification & Validation (V&V), Fidelity Assessment, Credibility Assessment, and the final compliance demonstration. These elements will be discussed in the following sections.

12.3 Phase 2a: FSM Build and Development

The RoCS project was provided with flight test data and a FLIGHTLAB FSM of the AW109 Trekker (F-AW109) by Leonardo Helicopter Division (LHD) with which to exercise aspects of the RCbS process. Flight data for trim, stability and response assessment were provided to the RoCS team for a range of test conditions, prior to any FSM analysis. Note that, in the formal RCbS process, the flight test data would be gathered in Phase 2, in conjunction with the development of the FSM and FS and following the development (incl. V&V) of the FTMS.

FSM Build. In accordance with the RCbS process, the FSM development phase should follow a structured approach with basic V&V building up from component to aircraft level. Physics-based modelling is considered an essential prerequisite to support extrapolation, as is the situation in this Case Study. The baseline FSM that formed the starting point for the RoCS activities was developed by LHD and was validated for up-and-away flight conditions, but not for low-speed conditions in proximity to the ground. The continued FSM development within the RoCS project was jointly aimed at the simulation of low-speed C&M, Category A take-off procedures and Dynamic Stability, the three Case Studies presented in this Book.

The baseline FSM features a rigid articulated blade-element main rotor. The tail rotor is modelled as a disk-type collective-only rotor (Bailey model), with aerodynamic properties originally tuned to level-flight pedal-to-yaw frequency response characteristics. The main rotor induced velocities are computed with a Peters-He finite-state inflow model (3-state as baseline), along with a source-image ground-effect model. The rotor aerofoil data are available in the form of lookup tables of the aerodynamic coefficients C_l, C_d and C_m as functions of angle of attack and Mach number. The blade airloads are computed in a quasi-unsteady fashion including unsteady circulatory effects from thin airfoil theory. The fuselage aerodynamic loads are computed at, and applied to, a single computational point. The fuselage and empennage force and moment coefficients are available as functions of angles of attack and sideslip, derived from model-scale wind tunnel test data.

The baseline FSM displayed several fidelity deficiencies in hover and low-speed flight (incl. hover in crosswind). The prediction of control positions and attitudes in trim, critical for this ACR, did not meet typical fidelity standards, e.g., those of CS-FSTD(H) [8]. In fact, predicting the interactional aerodynamic effects that occur on conventional main-tail rotor helicopter configurations in IGE hovers with winds from all directions is likely beyond the capability of current state-of-the-art simulation methods for real-time application. Figure 12.1 shows a typical comparison of tail rotor collective for a 20 kts wind around the azimuth, predicted from the F-AW109 FSM trim analysis and estimated from flight test measurements using the pace-car equivalent test method. While the comparisons are reasonably good for winds from 0° to180°, including the 90° critical azimuth, the data highlights fidelity deficiencies with winds from the 3rd and 4th quadrants. These are the regions where critical azimuths become related to HQ deficiencies. Evidence for such is provided by the large scatter in the flight data at the 220° and 270° conditions, particularly for the IGE cases, indicating the difficulty that the pilot has in maintaining the trim. The main and tail rotor wake modelling become particularly important in these conditions for achieving sufficient fidelity. Tail rotor VRS is likely to feature strong around the 270° condition while the interaction of the main rotor wake with the tail rotor is a likely contributor to the deficiency at 315°. A general handling issue in these quadrants, which does not necessarily lead to a limitation, is the loss of yaw ‘static stability’, where the pedal required to hold a new azimuth is in the opposite direction to the pedal input required to change the azimuth in the direction of the new azimuth (see e.g. [9]).

As stated, one of the shortcomings of the physical modelling lies in the approximation of the interactional aerodynamics. An attempt was made to improve the fidelity of the FSM to enable a sensible exercising of the RCbS process. The typical fuselage interference modelling (e.g. in FLIGHTLAB [10]) uses a lookup table that provides the aerodynamic force coefficients as functions of the angles of incidence

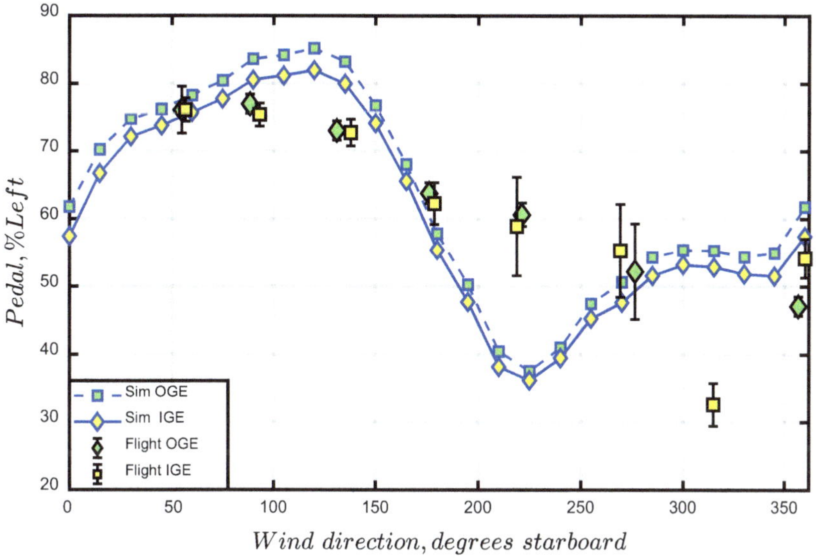

Fig. 12.1 Comparison of pedal position (tail rotor collective), as a function of wind azimuth angle, estimated from flight and predicted by the F-AW109; AW109T flight conditions, FLT140/142: wind $= 20 \pm 2.5$ knots, GW/σ $=$ 3065/3125 kg, Forward (FWD) CG, IGE/Out of Ground Effect (OGE) (IGE $=$ 35 ft AGL)

and sideslip. In the baseline FSM, a single Aerodynamic Computational Point (ACP) was used, disregarding the distribution of interference velocities and cross-sectional area of the fuselage. To account for these effects empirically, multiple ACPs have been specified by a set of locations along the length of the fuselage with a weighting defined by the local fuselage volume or area projected on the horizontal plane. The interference velocity vector used for table look-up is obtained through weighted averaging across the ACPs. The one-way look-up table interference modelling relies on the validity of the aerodynamic data table and the main rotor wake model. In this case, the data table was compiled from multiple wind tunnel experiments of different airframe configurations. The composite nature of the data raises some uncertainties considering the applicable range of incidence angles, but addressing these was beyond the scope of the RoCS project. Instead, aiming also to improve rotor performance modelling itself, effort was put into increasing the fidelity of the finite-state main rotor inflow model through ΔL-matrix augmentation derived from a Viscous Vortex Particle Method (VVPM) model, based on the method proposed in Refs. [11, 12]. It was found, however, that the number of inflow states required to approximate the VVPM inflow distribution was prohibitive for real-time piloted application and the effort was discontinued.

Another crucial aspect of the FSM for the low-speed controllability ACR is the fidelity of the tail rotor model. As noted above, the pedal control authority and the handling qualities in the yaw axis are heavily influenced by main rotor wake—tail rotor interference and tail rotor VRS conditions. The baseline disk tail rotor modelling can capture such effects only with a relatively low level of fidelity. Part of the off-line analyses and FS testing was, therefore, performed with a blade element tail rotor model with finite-state inflow. The off-line pedal trim results improved in comparison to flight test in winds from around 11 o'clock, but in the conditions tested in the FS the pilot did not report a significant influence on the perceived handling qualities, although the findings on this topic reported in this Book are not considered conclusive. Note, however, that for the following sections, the precise level of fidelity achieved is not considered critical to the objective of exercising the RCbS process.

FSM Verification. In relation to the FSM and associated analysis routines, the verification process aims to (1) verify the correctness of the numerical implementation and, (2) estimate the numerical error and associated uncertainty. The second step, commonly referred to as *solution verification*, is dependent on the application or problem of interest. The established practices outlined in, e.g. Ref. [13], do not necessarily translate directly to the multi-dimensional and multi-physics problem of rotorcraft flight simulation. Nevertheless, particularly when solutions are constrained to run in real-time, where both the time step itself and the iterative convergence criteria for each step may need to be relaxed, the impact of such accommodations should be assessed. A pragmatic approach to considering the effect of discretization [14] is to consider the numerical uncertainty u_{num} as epistemic, i.e., an interval without associated probability distribution, where the bounds on numerical uncertainty for a given output parameter of interest are defined equal to $\pm$ the magnitude of the error relative to the *fine grid* solution.

In the low-speed controllability case, the analysis may consider, e.g., the prediction of the pedal trim position or the peak yaw rate response to a pedal input. The discretisation parameters for a typical implementation may include the solution time step, the number of rotorblade aerodynamic segments, the number of inflow states, or the number or maximum age of particles in case of Vortex Particle Method analysis. Figure 12.2 provides an example of the convergence of the predicted hover-in-wind trim pedal position as a function of such discretisation parameters (multiple colour lines), where the arrow indicates the associated numerical error of the baseline model with nominal discretisation with respect to the *fine grid* solution (in blue). Note that, for this case, the corresponding numerical (output) uncertainty of $\pm$ 3.2% is comparatively large and a refinement of the discretisation of the nominal model would be required, subject to constraints imposed for real-time piloted simulation, if applicable.

FSM Validation. The FSM validation process, stipulated for instance in CS-FSTD(H), consists of a comparison of the (tuned) simulation prediction against flight test data and establishing criteria for the acceptable level of mismatch. However, to support the credible extrapolation of the simulation predictions to conditions for which no flight test data are available, it is important to develop an understanding of the uncertainties present in both the simulation predictions and the flight test, or, more generally, the 'referent' data. Chapters 6 and 9 cover this topic in more depth. A specific goal of the validation exercise is to quantify the range within which the

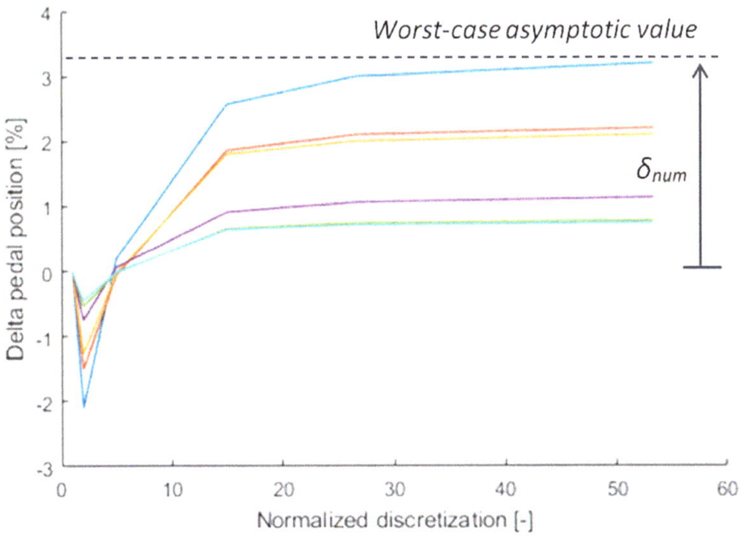

Fig. 12.2 Discretisation convergence study to establish DoV numerical uncertainty, u_{num}, in trim pedal position considering the various individual and combined discretisation (different colours) present in the simulation model

model error lies ($\delta_{model} \in \delta_{val} \pm u_{val}$, Eq. 6.7); i.e. within range of the validation error δ_{val} spanned by the validation uncertainty (u_{val}). This range can then be used in Phase 3 of the RCbS process to establish the total simulation prediction error, δ_p, or its uncertainty in conditions other than those for which validation data are available. Depending on the conditions, this will require either interpolation (in the DoV) or extrapolation (in the DoE) of the model error and uncertainty. The numerical and input uncertainties can be reassessed directly at the prediction conditions. Note that the validation uncertainty includes the measurement (referent) uncertainty u_r, which can be carried over to the prediction conditions in the DoE, thus impacting confidence in the predictions at the extrapolated condition. An important point here is that this means that large measurement uncertainty in the DoV will be reflected as prediction uncertainty at the extrapolation conditions. Reductions in the uncertainty of the measurement can be achieved in various ways, including by using different or complementary instrumentation, but also by increasing the number of samples, e.g., by repeating test points or increasing the measurement period. Note that the latter does not necessarily reduce the uncertainty, but does increase the confidence in the characterisation of that uncertainty.

Figure 12.3 provides an example validation exercise based on a trim assessment of the pedal position for an out-of-ground effect hover in 20 ± 2.5 knots wind from 90° ± 10° azimuth at maximum take-off gross weight, as a function of density altitude. The data have been obtained through certification flight test activity on the AW109 Trekker helicopter. The indicated ranges for wind speed and direction reflect the differences between the test conditions, not the variation for a given trim condition.

The error bars on the flight test data in the upper chart reflect the uncertainty in the measurement. In the absence of information on the uncertainty related to the instrumentation, only the process uncertainty associated with the test procedure (typically dominant) is accounted for here. That is, the uncertainty in the mean value computed from multiple trim recordings, as described by a Student's *t*-distribution [15]. The error bars around the simulation result represent the simulation total validation uncertainty ku_{val} with coverage factor $k = 2$ (appropriate for a situation involving only aleatoric uncertainty, see Chaps. 6 and 9). The lower chart shows the validation error δ_{val} (square marker), with the error bars indicating the validation uncertainty u_{val}. The model-form error δ_{model}, derived from Eq. (6.7), must fall within the indicated range.

As discussed in Chap.'s 6 and 9, the effects of the input uncertainty u_{inp} can be established through a Monte-Carlo analysis considering the uncertainties in the model input parameters. These uncertainties will vary between validation conditions, where much is known about the test aircraft, and predictions beyond the DoV, where, e.g., operational variations of parameters across the fleet may need to be considered. To reduce the scope of the input uncertainty quantification outside of the DoV, conservative worst-case assumptions can be made for certain input parameters, such as the control rigging or minimum specification engine power, rather than specifying an interval or probability distribution. This approach is only possible in cases where it

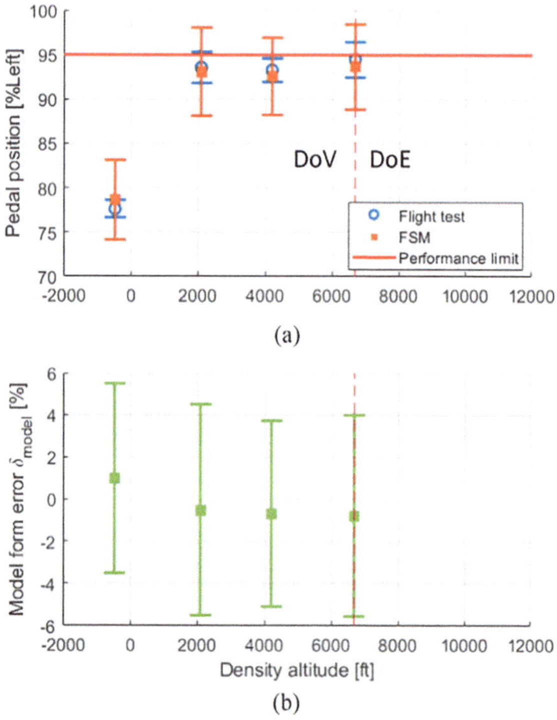

Fig. 12.3 Validation of trim pedal position prediction for OGE hover in 20 ± 2.5 knots wind from 90° ± 10° azimuth at maximum take-off gross weight as a function of density altitude; **a** pedal position (%), **b** model-form error (%)

can truly be argued that the worst-case input parameter value leads to a worst-case result in terms of all output quantities of interest.

It can be argued based on Fig. 12.3 that it is difficult, with the test data available for this aircraft below 7000 ft, to justify a trend for extrapolation of the model error and associated uncertainty to the maximum take-off and landing altitude. In fact, besides density altitude, also the gross weight of the aircraft varies. Additional test points at intermediate altitudes, and constant referred gross weight (GW/σ), are desired to establish a credible trend and tighten the associated confidence intervals. Beyond the availability of test data, an alternative approach to the one taken in Fig. 12.3 would be to compute (and flight test), at the 90° azimuth, the wind speed that requires a trim tail rotor collective position of, say, 95%, down to a minimum wind speed of 17 knots and adjusting the take-off gross weight as needed. This would ensure a greater consistency in the tail rotor state (at least in terms of thrust), amongst the validation conditions used to define an extrapolation to the maximum take-off and landing altitude.

Note that the process of establishing the validation uncertainty leads to a reduced emphasis on model tuning and the questionable nature thereof in terms of extrapolation. In fact, the input uncertainty u_{num} should encompass the range of feasible tuning parameters. The mean values of the output probability distributions can then be used to characterise the fidelity of the simulation relative to the referent through established or applicant-defined fidelity metrics, as will be discussed in the next section.

It may be tempting to forego the uncertainty quantification effort in favour of a conservative approach, in which the error relative to flight test is accepted, as long as it is conservative (even if this comes at the expense of performance). However, even with this approach, it must be shown that the uncertainty of the prediction is not larger than the perceived conservativeness of the nominal FSM.

FSM Predictive Fidelity. Fidelity metrics are defined to provide a quantification of the level of fidelity such that criteria can be agreed for what constitutes sufficient fidelity. The key challenge is defining what is considered sufficient fidelity for a given ACR. Moreover, what may be defined as acceptable in the validation phase, may need to be reconsidered in the credibility or compliance demonstration phase based on the proximity to non-compliance with the requirements of the ACR. Where possible, it is desirable to define acceptable error tolerances, in time or frequency domains, in direct relation to the requirements of the ACR. Otherwise, it may be possible to leverage existing standards like CS-FSTD(H), if HQ and response aspects are relevant to the ACR.

Whereas the analysis to establish the validation uncertainty, with the purpose of underpinning an extrapolation to compliance demonstration conditions, may be limited to the primary parameters of interest (i.e., those directly related to the requirements of the ACR), the fidelity assessment must consider all relevant degrees of freedom. However, as in CS-FSTD(H), the acceptable error tolerance will vary between parameters depending on the ACR. Given the fact that this ACR revolves around controllability, the time-domain fidelity metrics and tolerances of CS-FSTD(H) may be applied:

- Trim:
  - Control positions: ± 5%
  - Pitch attitude: ± 1.5°
  - Bank angle: ± 2°
  - Engine torque: ± 3%
- Control response (at critical azimuth):
  - Yaw rate: ± 10% or ± 2°/s

The tolerance of 5% on trim pedal position may be rather large for the critical azimuth where the trim margin is smallest and the applicant will want to achieve the maximum performance. The applicant may, therefore, aim for a higher accuracy from the outset for that particular trim parameter.

Frequency-domain assessments using, e.g., bandwidth and phase delay, or integrated cost functions of the magnitude and phase of the frequency response, provide a meaningful complement to time-domain analyses [16, 17]. These frequency-domain methods are particularly suitable for application to ACRs with a handling qualities component, especially those involving small-amplitude perturbations from trim, as they are essentially linear methods. They are viewed to be less useful for the nonlinear, unsteady conditions of some ACRs. In practice, however, such test data may be difficult to obtain in the most relevant conditions, e.g., in quartering flight close to the ground.

Figure 12.4 shows the frequency-domain comparison between flight and the F-AW109 FSM for the hover pedal to yaw angle response (derived from the pedal to yaw rate response). The grayscale on the flight test data, and the 95% confidence interval, are based on the coherence and the random error function [17]. The flight test data quality is such that the phase-limited bandwidth is undefined, but an integrated cost function of the frequency response, J_{ave}, taking into account the coherence function, can nevertheless be obtained.

Reference [16] proposes a guideline of $J_{ave} \leq 100$, where for a Multi-Input/Multi-Output system of n_{TF} transfer functions, J_{ave} is defined as:

$$J_{ave} = \sum_{l=1}^{n_{TF}} \left\{ \frac{20}{n_\omega} \sum_{\omega_1}^{\omega_{n_\omega}} \left(W_\gamma \left[W_g \left(|\hat{T}_c| - |T| \right)^2 + W_p \left(\angle\hat{T}_c - \angle T_c \right)^2 \right] \right) \right\}_l \frac{1}{n_{TF}} \tag{12.1}$$

In Eq. (12.1), W_γ, W_g and W_p are frequency (ω) dependent weighting functions on coherence γ, and the magnitude $|T|$ and phase $\angle T$ of the transfer function T. Given the weighting function conventions used in Ref. [16] and a frequency range between 0.3 and 10 rad/s, the cost function for the singular on-axis response shown in Fig. 12.4 is $J = 97.3$.

Finally, metrics are also required to establish the fidelity achieved in the flight simulator. For this, an applicant may propose one or more subjective rating scales such as the SFR described in Ref. [18] and/or the Cooper-Harper Handling Qualities Rating (HQR) [19]. In both cases, it will be important that the FTMs are performed

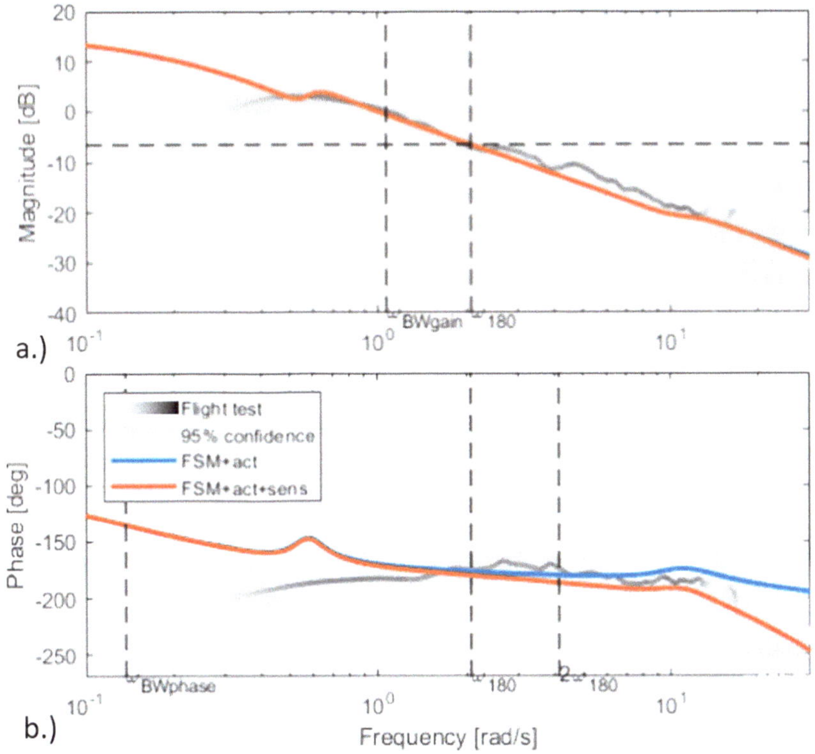

Fig. 12.4 OGE hover, w/o wind, pedal to yaw angle frequency response comparison between flight test, FSM with actuator modelling (blue line), and FSM with both actuator and sensor modelling (red line); **a** gain (dB) and **b** phase (°)

and assessed consistently between flight and simulation, preferably by the same pilots. In addition, objective metrics, e.g., based on control activity or eye tracking, may be included to complement and verify the subjective assessment.

12.4 Phase 2b: FS Build and Development

The RCbS process prescribes a rigorous FS build and development process following the same steps as described for the FSM. In most cases the FS is likely pre-existent, in which case documentation of past efforts may be partly relied upon, assuming configuration management and record-keeping are in order. Nevertheless, dedicated V&V is expected to be required. As discussed in Chap. 11, within the RoCS project, the FS trials presented in this Book were performed using the University of Liverpool's (UoL) HELIFLIGHT-R simulator for the FS validation and compliance demonstration (Fig. 12.5, Ref. [20]). The simulator has a 230° × 70° field of view, with a

Fig. 12.5 Cockpit view of X-Wind Hover (XWH) low-speed controllability ACR simulator set-up in UoL's HELIFLIGHT-R simulator [21]

4-axis force feedback control loading system and a 6 degree of freedom electric motion platform (see Chap. 11).

Both FTMs were explored; the pace-car equivalent (PCE) FTM, which replicates the typical flight test practice (Appendix 1), and the cross-wind hover (XWH) FTM, which is considered more representative of the operational environment in which low-speed controllability issues may be encountered (Appendix 2). The aim was, in part, to compare qualitatively the FTMs in terms of their efficacy in identifying low-speed handling quality deficiencies. Moreover, as mentioned previously, a one-to-one piloted validation requires that the evaluation is performed using the same FTM in the FS as in flight.

The XWH FTM visual references for desired (±2 ft) and adequate (±4 ft) height performance assessment were provided using blue and yellow hover boards and a red reference height ball at 45° intervals around the azimuth, as shown in Fig. 12.5. Desired (± 3 ft) and adequate (± 6 ft) plan position performance references were provided by cones placed at each 45° azimuth. The relative wind condition could be set using the FS' Environment feature to give the required test wind velocity at any given 'visual' azimuth. For the PCE FTM, the visual database developed for the companion Dynamic Stability FTM (Ref. [21], and Chap. 14), was used, consisting of a runway with additional visual content (buildings) alongside it for height references. The head-down flight display used in AW109 flight testing was emulated to provide information to the co-pilot on ground speed and ground track angle for assessing the 'validity' of a test point. The Evaluation Pilot (EP) flew the manoeuvre using

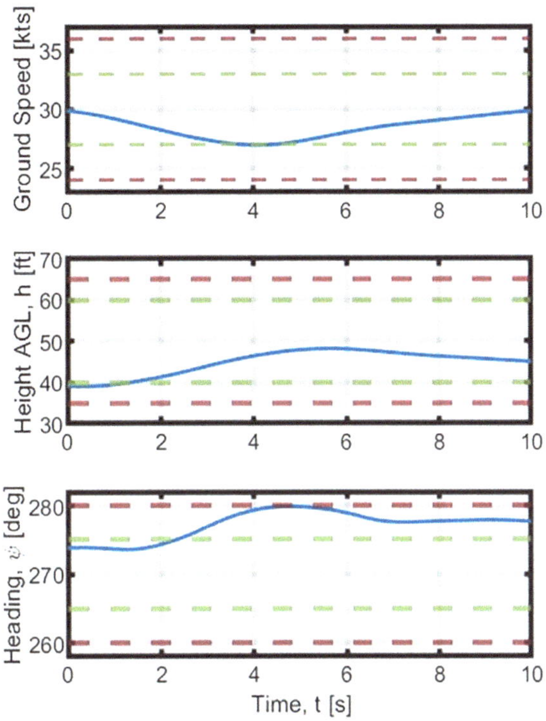

Fig. 12.6 PCE FTM task performance for 30 kts, 270° wind azimuth within the DoV

outside-world visual references and was aided by the co-pilot in the left-hand seat calling out ground speed.

HELIFLIGHT-R's Vestibular Motion Cueing feature was tuned during the pre-trial work-up. Previous work at UoL [22, 23] has shown that careful selection of the parameters in the motion drive algorithms, gain and break frequency in translational and rotational axes is required to ensure sufficient cueing for a given task, including minimal adverse cues. The values tuned for the XWH/PCE FTMs are provided in Appendix 3.

As an example, Fig. 12.6 shows the task performance achieved for the PCE FTM at a 30 kts, 270° wind azimuth condition in the nominal DoV (i.e., wind from 9 o'clock), which was part of an airspeed sweep starting from a hover. The FSM had already been validated against flight test data at 90° wind azimuth (i.e. wind from 3 o'clock) where pedal authority was the limiting factor (Fig. 12.1). However, at 270° the wind speed limits are expected to be defined by handling qualities deficiencies, so the aim here was to obtain feedback from the EP on the handling qualities and the fidelity of the FS features. Recall that, from Fig. 12.1, the FSM is known to have significant fidelity deficiencies at this condition, so tests with the baseline model were not expected to be realistic. The green and red dashed lines in Fig. 12.6 represent the boundaries of the desired and adequate FTM performance standards. Whilst an HQR was not awarded for this test point, the task performance was within desired

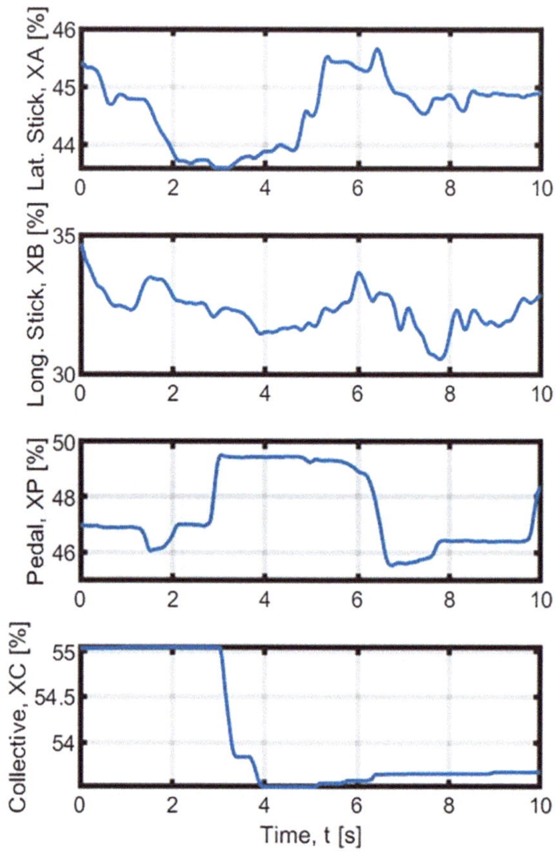

Fig. 12.7 PCE FTM control activity for 30 kts, 270° wind azimuth condition, within the DoV

performance for height and speed. There was an initial offset in the heading datum, but the EP could complete a heading change in this condition, in-line with the ACR, without encroaching on control limits or encountering controllability limitations. Pilot control activity is shown in Fig. 12.7 for the 30 kts condition, indicating that no control limits are approached at this condition, and pilot compensation was reported as minimal during the on-condition hold period. For the same azimuth, the ground speed was increased until a value of 40 kts was achieved, again without encroaching on control limits or requiring excessive pilot compensation for handling deficiencies. To re-iterate, this condition was not formally evaluated via the award of an HQR.

At 42 kts the task was abandoned as it was not possible to maintain FTM performance parameters in height and heading. The EP reported strong side force cues considered to be due to yaw, and heave due to sideslip velocity couplings, commenting that neither of these phenomena were expected dynamics of the Trekker. The two deficiencies made heading control and height control difficult at high wind speeds, and created a handling qualities limit, as may be expected at this wind azimuth.

Note that, although the FS testing served its purpose in terms of exercising the RCbS process and exploring the FS set-up, it should be clear for those that are familiar with helicopter low-speed controllability testing, or the AW109 Trekker in particular, that a wind speed of 40 kts at a relative direction of 270° is excessive for this generation of helicopter. Also, as already discussed, the FSM employed here was not able to reproduce the (pedal-related) handling qualities limitations experienced on the real aircraft in the azimuth range between 180° and 270°, likely related to the fidelity deficiencies in the main rotor wake interactions with the fuselage and ground surface, and tail rotor VRS.

12.5 Phase 3: Credibility Assessment and Compliance Demonstration

FSM Credibility Assessment. Ultimately, it is the task of the applicant to convince not only the certification verification engineer (CVE) in their Organisation, but of course the certification authority, that the simulation results are credible. In the absence of test data at the certification conditions, it is impossible to provide mathematical proof, but efforts can be made to quantify the confidence in the predictions. As emphasised throughout this Book, beyond the quantifiable metrics, there are other factors that influence the credibility, such as the expertise and experience of the engineers involved and the rigour applied to the simulation development and its application as a whole. The following discussion will be limited to the uncertainty quantification of the prediction in the DoE as it relates to the FSM.

Assuming, for simplicity, independence of its components, the prediction uncertainty in the DoE, u_p, can be written as:

$$u_p = \left(\sqrt{u_{num}^2 + u_{inp}^2 + u_{model}^2}\right) \tag{12.2}$$

where the model-form uncertainty at the prediction condition, u_{model}, can be estimated through extrapolation of u_{val}, Eq. (6.6).

In the current example, δ_{val} in the DoV is of similar order of magnitude as u_{val}, as evident from Fig. 12.3. The prediction error at the extrapolated condition can then be bounded through the expression:

$$\delta_p \in \delta_{val} \pm u_p \tag{12.3}$$

Here, lacking a referent, δ_{val} represents the extrapolation of the validation error, computed in the DoV, into the DoE. The extrapolated error, computed from an FSM that has been updated, tuned and validated in the DoV, and under the condition that the validation uncertainty is of similar or smaller magnitude compared to the error, is used in the DoE as a 'correction' to the simulation prediction. Such corrections are particularly important if the FSM has been shown to have a non-conservative bias in the DoV which cannot be otherwise reduced by model tuning and updating within practicable, physics-based, limitations.

The DoE prediction uncertainty u_p is similarly obtained through extrapolation, as shown in Fig. 12.8. Herein, the dashed red lines indicate the 95% confidence bounds associated with the extrapolation of the linear regression model fit. As a conservative approach, instead of adopting the confidence bounds directly, the model-form uncertainty and the prediction error at 12,000 ft are identified by the green circles in Fig. 12.8, equal to 3.6% and − 6.2% of the total pedal travel, respectively. The conservative end for the extrapolated validation error is on the negative side since this represents the case where the model predicts a lower pedal position than the measurement (expected truth). Note that the extrapolation in Fig. 12.8 is one-dimensional in density altitude, whereas a multi-dimensional regression fit would be more appropriate.

Figure 12.9 presents the resulting trim pedal margin prediction in the DoE. The simulation prediction at 12,000 ft density altitude includes the uncertainty bounds accounting for extrapolated model-form uncertainty, as well as the input and numerical uncertainties at the prediction conditions, as per Eq. 12.2. It is clear from Fig. 12.9 that the prediction exceeds the performance limit for a 'noticeable yaw response' and that the uncertainty around the simulation further reduces the performance margin. This example highlights a case where efforts should be made to further reduce both the error and the uncertainties to a level that the price paid for the use of simulation in terms of performance margin is acceptable.

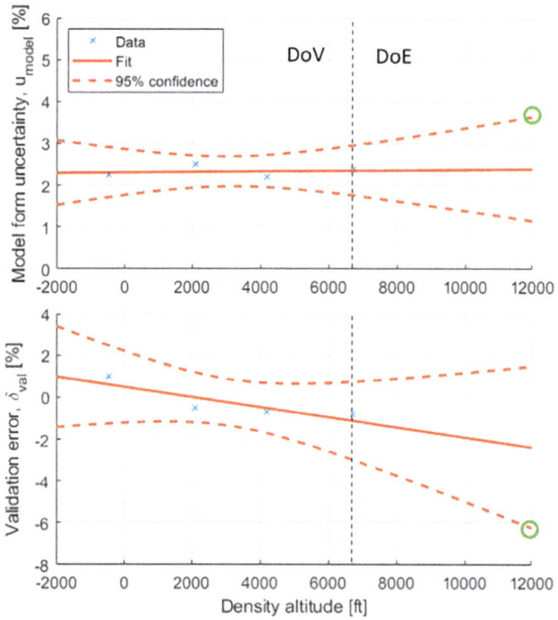

Fig. 12.8 Linear regression fit of simulation model-form uncertainty and validation error (including the extrapolated validation error), with 95% confidence bounds, for an OGE hover in 20 ± 2.5 kts wind from 90° ± 10°. Negative validation error represents a non-conservative prediction

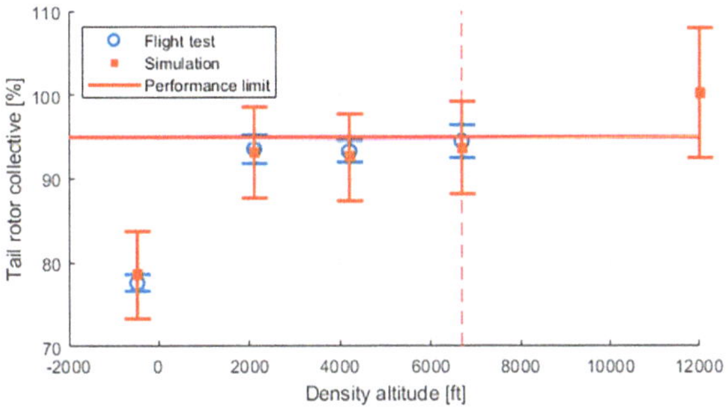

Fig. 12.9 Extrapolation of trim tail rotor collective up to 12,000 ft density altitude, for an OGE hover in 20 ± 2.5 kts wind from 90° ± 10°. The blue and red vertical solid lines indicate the flight test and simulation prediction uncertainties, respectively

The credibility analysis described above has been limited to an off-line trim pedal margin assessment with wind from the critical azimuth (90°). Typically, as discussed earlier, there are also envelope limitations related to handling qualities. For the AW109 Trekker, the wind envelope is limited by handling qualities at a relative wind direction between 160° and 260°. The simulation-based compliance demonstration for such limitations requires piloted simulation in a suitable FS, as will be discussed in the following section.

DoE FS Testing. This section contains preliminary results from piloted simulations conducted at The University of Liverpool in which several EPs flew a mock certification compliance demonstration simulation trial for the low-speed C&M ACR.

The FSM used for the piloted trials corresponded to the configuration described in Sect. 12.3 without further modification or variation. A more rigorous approach would involve a similar FSM uncertainty quantification exercise as described above, and with a focus on relevant objective handling qualities metrics. In this way, the FS trials could be set up to reflect an FSM configuration that falls on the conservative side of the uncertain range of predicted handling qualities.

FTM Comparison. The compliance demonstration phase of the FS trials was performed using the XWH FTM, but the flight condition was also assessed using the PCE FTM to facilitate comparison of the two methodologies. Figure 12.10 shows the task performance over the 10 s 'stabilised' phase, for both FTMs for a 30 kts, 240° wind azimuth case within the DoE. Note that the green and red dashed lines represent the performance standards, but in plot (b) the dashed red line indicates adequate performance for the XWH FTM, and the dotted red line indicates the adequate performance standard for the PCE FTM.

In the XWH FTM, one of the EPs reported there was "beyond extensive" pilot compensation to capture the required wind azimuth angle and once 'on condition' there was considerable to extensive pilot compensation required to achieve adequate task performance. The results shown are from an exploratory test, when control compensation required to capture the desired heading was not accounted for in the award of the HQR for the FTM assessment presented here. However, control activity (Fig. 12.11) did provide insight into the handling qualities deficiencies of the simulated aircraft, primarily due to the lateral-translation cues due to yaw also identified in the PCE FTM. Plan position was 'adequate' and there was a heading bias that

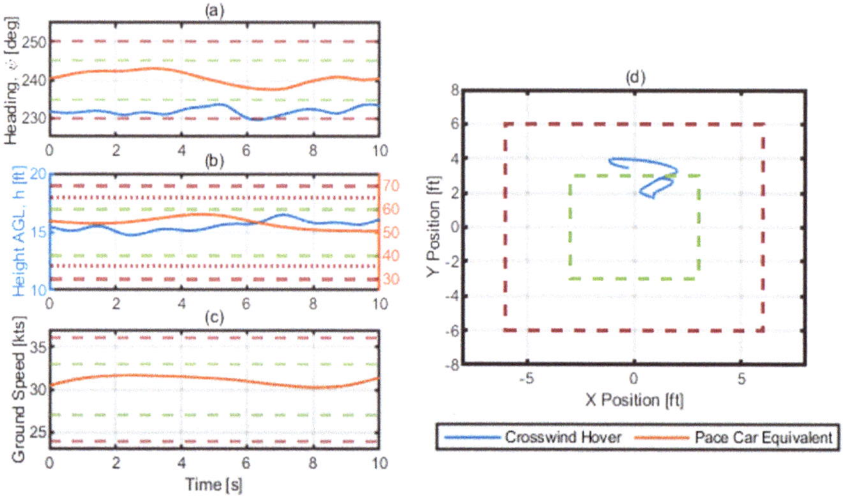

Fig. 12.10 XWH and PCE FTM performance for 30 kts wind at 240° azimuth in the DoE

shows only adequate performance achieved in heading, which the pilot commented was due to a high angle of bank in the trim condition. Conversely, in the PCE, both the heading and ground speed remained within desired performance bounds throughout the required 10 s on-condition hold period.

The pilot control activity during the heading hold phase shown in Fig. 12.11 reveals oscillatory inputs in all 4 control axes for the XWH manoeuvre (pilot compensation 'considerable'), which are not present in the PCE. These oscillatory inputs appear to be driven by higher piloting gains forced by the performance standards of the XWH which, relative to the PCE, are tighter; a 3 kts ground speed (desired performance in the PCE) would take the pilot out of desired performance in the XWH FTM in approximately 0.6 s, which a pilot would struggle to respond to quickly enough to correct for. As a result, the pilot had to apply larger, more rapid, corrective inputs to maintain the given performance standards.

Generally, the perceived handling qualities in the XWH and PCE FTMs were not equivalent even though the physics of the simulated aerodynamics are theoretically identical. The different piloting strategies highlighted different characteristics and deficiencies in the FSM, the FS and hence the perceived handling qualities. The XWH, with the associated performance targets specified, was characterised by higher piloting gains. The disparities between the two FTMs highlight a situation where performance may be degraded and result in a poorer HQR in a more representative task than the existing compliance demonstration methodology.

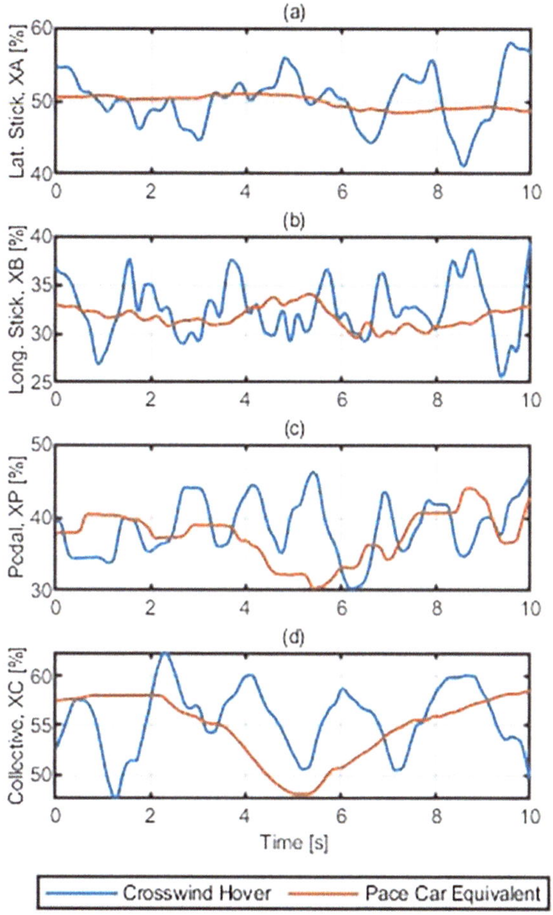

Fig. 12.11 XWH and PCE EP control activity for 30 kts wind at 240° azimuth in the DoE

12.6 Concluding Remarks and Recommendations

This Chapter has reported on an exercise of the RCbS process and presented results from the Case Study on the low-speed controllability and manoeuvrability ACR as expressed in EASA certification specifications CS-27 and 29. The Case Study provided a useful basis for exploring the practical implementation of the RCbS process and its associated benefits and limitations. FSM V&V and credibility assessment efforts focused on the prediction of the trim pedal margin limit for which the current FSM is considered suitable.

Conclusions and associated recommendations from the Case Study are as follows:

(i). The adopted FSM displayed sufficient fidelity for the evaluation of trim control margins for the C&M assessment at the 90° wind critical azimuth.

(ii). In contrast, the FSM was not able to reproduce the trim pedal in the 3rd and 4th wind azimuth quadrants, and consequential low-speed handling quality deficiencies of the real aircraft; advances in the state of the art are required for this ACR to be addressed fully through real-time simulation. Achieving sufficient fidelity for both off-line FSM and piloted FS testing requires improved wake and interactional aerodynamics modelling for these flight conditions.

(iii). Using the FSM outside of its DoV inherently introduces additional uncertainty. The quantification of this uncertainty, as presented in this Case Study, is one of the key elements of the simulation credibility assessment. As shown, the prediction uncertainty unavoidably comes at the expense of performance margin.

(iv). The case study has emphasised that the uncertainty-centric approach taken in the RCbS V&V and credibility assessment phases puts a new perspective on FSM fidelity and model updating, especially in areas of the flight envelope where no test data are available (e.g. in the DoE).

(v). Two FTMs (XWH and PCE) were developed for a preliminary piloted assessment of fidelity and handling qualities related to the C&M ACR. The FTMs did not produce equivalent HQ ratings, highlighting different perceived deficiencies and differences in piloting control strategies. The XWH is considered more representative of the operational scenario relevant to this ACR but further investigations are required, particularly relating to cueing and performance standards, before a final task definition can be recommended.

Appendix 1: PACE-Car Equivalent (PCE) FTM Description

Mission	Civil transport
Critical HQs	Yaw attitude quickness and bandwidth Lateral-directional stability Pitch and Roll bandwidth Cross-couplings: pitch/roll, roll/pitch, collective/yaw, collective/pitch
Objectives	• Assess handling qualities in 'hover' under crosswind conditions • Check ability to perform heading transitions in crosswind conditions • Assess the effect of vestibular motion cueing on pilot perception of simulation fidelity

(continued)

(continued)

Manoeuvre description	The FTM begins with the aircraft in a stabilised hover, 50 ft AGL, at one end of the runway on a heading of ψ_W, over the runway centreline. The EP will initiate a transition along the centreline of the runway, accelerating to a ground speed of V_W. Once the aircraft has reached V_W, the EP must maintain ground speed, heading and altitude within specified performance standards, for a period of 10 s. A 10s timer can be initiated by pressing the cyclic grip button; the EP must not initiate the timer before reaching V_W. The EP will be cued once 10 s has elapsed, the pilot must adjust heading to exceed $\psi_W + 5°$ whilst maintaining runway centreline, altitude, and ground speed, and then adjust heading to exceed $\psi_W - 5°$, again maintaining other performance standards. The task is complete once the pilot has attained $\psi_C \pm 5°$ following heading hold		
Test course description	Standard runway in an open area. Testing will be conducted in VMC		
Test variations	Condition	Mass	Pressure altitude
		3175 kg Increments up to MTOWkg	745 ft 10,000 ft
Ratings scales	1. Cooper-Harper Handling Qualities Rating Scale 2. Simulation Fidelity Rating Scale		
Performance standards		Desired (d)	Adequate (a)
	Maintain ground speed, δV_W	± 3 kts	± 6 kts
	Maintain altitude δh	± 10 ft	± 15 ft
	Maintain heading $\delta \psi_W$	± 5°	± 10°

(continued)

(continued)

GENESYS
White line 'driven' by model output of groundspeed and ground track angle. White circles at 10,20 knots

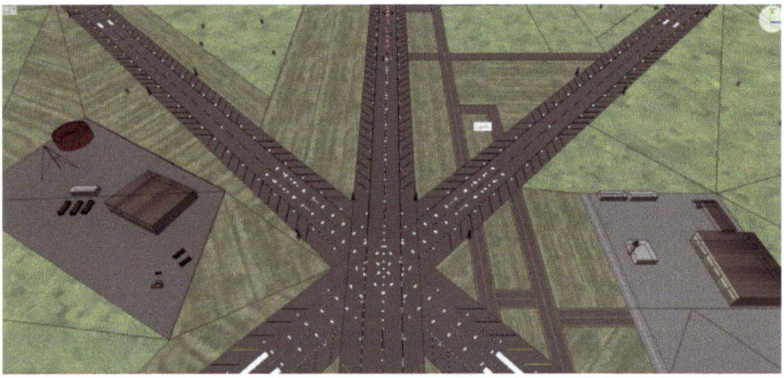

Appendix 2: Cross-Wind Hover (XWH) FTM Description

Mission	Civil Transport		
Critical HQs	Yaw attitude quickness and bandwidth Lateral-directional stability Pitch and Roll bandwidth Cross-couplings: pitch/roll, roll/pitch, collective/yaw, collective/pitch		
Objectives	• Check ability to perform precise hover control • Check ability to perform heading transitions from hover to hover under wind conditions • Assess the effect of vestibular motion cueing on pilot perception of simulation fidelity • Assess sufficiency of visual cueing for the task		
Manoeuvre description	The EP will be released trimmed in a hover above the helipad, at a skid height of $h =$ 10 ft, on a nominal heading, ψ, 45° from the critical azimuth. The EP will be expected to perform a 45° left-hand turn onto the critical azimuth and maintain a 10 s hover once stabilised. The EP will press the cyclic grip trigger button to indicate they have achieved a stabilised hover at the new heading; this initiates an audio cue for the start and end of the 10 s timer. Following this, the EP will attempt + 5° and − 5° heading change from the critical azimuth while maintaining plan position and hover height. Performance standards will not be assessed during the initial azimuth capture but EP commentary for the transition phase will be recorded		
Test course description	Helipad surrounded by hover boards indicating lateral position and height performance standards on each heading. Cones will also be provided to indicate longitudinal position performance bounds for each hoverboard. Testing will be conducted in VMC		
Test variations	Condition	Mass	Pressure altitude
		3115 kg Increments to MTOWkg	745 ft 10,000 ft
Ratings scales (Max. 2 per test point)	1. Cooper-Harper Handling Qualities Rating Scale 2. Motion Fidelity Rating Scale		
Performance standards		Desired (d)	Adequate (a)
	Maintain lateral/ longitudinal position	± 3 ft	± 6 ft
	Maintain hover height δh	± 2 ft	± 4 ft
	Maintain heading $\delta\psi$	± 5°	± 10°

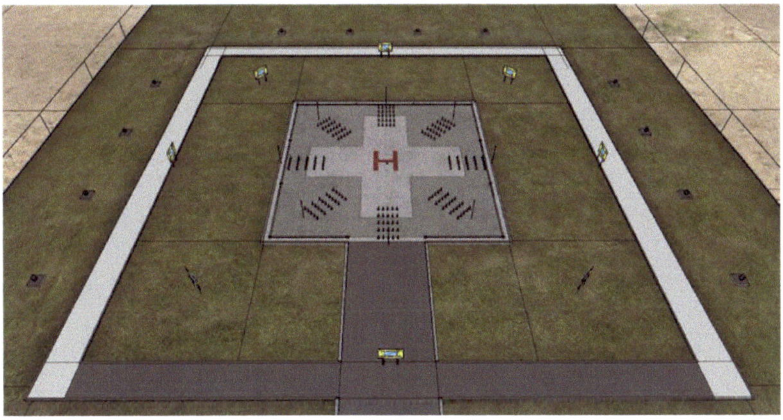

Appendix 3: HELIFLIGHT-R Motion Drive Law Parameters Used in the PCE and XWH FTMs

Simulator platform movements are determined by the Motion Drive Algorithm (MDA) that scale, limit and filter the signals from the FSM to generate Vestibular Motion Cueing commands. An example of the third-order filter used for the yaw axis MDA is given in Eq. (12.4) which scales and filters the FSM yaw acceleration, $\ddot{\psi}$, converting it into a commanded motion platform yaw acceleration, $\ddot{\psi}_s$. The parameters k_ψ and $\omega_{hp\psi}$ are the yaw high pass (*hp*) filter gain and break-frequency coefficients respectively, which are 'tuned' for a given FTM. Similar filters are used in other rotational (ϕ, θ) and the translational axes (x, y, z). Table 12.1 lists the MDA parameters used in the XWH and PCE FTMs.

$$\frac{\ddot{\psi}_s}{\ddot{\psi}}(s) = k_\psi HP_\psi(s) = k_\psi\left(\frac{s^2}{s^2 + 2\zeta_{hp\psi}\omega_{hp\psi}s + \omega_{hp\theta}^2}\right)\left(\frac{s}{s + \omega_{b\psi}}\right) \tag{12.4}$$

Table 12.1 XWH and PCE FTM MDA parameters

MDW	Surge		Sway		Heave		Roll		Pitch		Yaw	
	k_x	ω_{hpx}	k_y	ω_{hpy}	k_z	ω_{hpz}	k_Φ	$\omega_{hp\Phi}$	k_Θ	$\omega_{hp\Theta}$	k_Ψ	$\omega_{hp\Psi}$
XWH and PCE	0.5	1.0	0.4	1.0	0.55	3	0.28	0.4	0.2	0.4	0.6	0.4

References

1. EASA (2018) Certification Specifications and Acceptable Means of Compliance for Small Rotorcraft CS-27/Amendment 6
2. EASA (2018) Certification Specifications and Acceptable Means of Compliance for Large Rotorcraft CS-29/Amendment 5
3. FAA (2008, September) AC 29–2C—Certification of Transport Category Rotorcraft
4. He C, Lee CS, Chen W (1999, May) Finite state induced flow model in vortex ring state. In: American Helicopter Society 55th Annual Forum, Montreal, Canada
5. Johnson W (2005, December) Model for vortex ring state influence on rotorcraft flight dynamics. NASA/TP-2005–213477
6. Peters DA, He C (2006) Modification of mass-flow parameter to allow smooth transition between helicopter and windmill states. J Am Helicop Soc 51(3):275–278. https://doi.org/10.4050/1.3092888
7. U.S. Army Aviation Engineering Directorate (2000, March) ADS-33E—Aeronautical Design Standard Performance Specification: Handling Qualities Requirements for Military Rotorcraft
8. European Aviation Safety Agency (EASA) (2012, June) Certification Specifications for Helicopter Flight Simulation Training Devices CS-FSTD(H)
9. Anon (2019) Unanticipated Left Yaw, Airbus Safety Information Notice 3297-S-00. www.airbushelicopters.com/techpub/
10. DuVal RW, He C (2018) Validation of the FLIGHTLAB virtual engineering toolset. Aeronaut J 122(1250):519–555. https://doi.org/10.1017/aer.2018.12
11. He C, Syal M, Tischler MB, Juhasz O (2017) State-space inflow model identification from viscous vortex particle method for advanced rotorcraft configurations. In: American Helicopter Society 73rd Annual Forum, Fort Worth, Texas, USA
12. He C, Zhao J (2008, April) A real-time finite state induced flow model augmented with high fidelity viscous vortex particle simulation. In: American Helicopter Society 64th Annual Forum, Montreal, Canada.
13. ASME (2009) Standard for verification and validation in computational fluid dynamics and heat transfer: ASME V&V 20-2009
14. Roy CJ, Oberkampf WL (2010) A complete framework for verification, validation, and uncertainty quantification in scientific computing. In: 48th AIAA Aerospace Sciences Meeting, Orlando, Florida, USA. https://doi.org/10.1016/j.cma.2011.03.016
15. Whiting NW, Roy CJ, Duque E, Lawrence S, Oberkampf WL (2023) Assessment of model validation, calibration, and prediction approaches in the presence of uncertainty. ASME J Verif Valid Uncertain Quant. https://doi.org/10.1115/1.4056285
16. Pavel M, Tischler M, White MD, Stroosma O, Jones M, Miller D, Myrand-Lapierre V, Nadeau-Beaulieu M, Taghizad A (2021) Simulation fidelity assessment for rotorcraft methods and metrics: sketches from the work of NATO AVT-296. In: 77th Annual Vertical Flight Society Forum, Virtual, USA
17. Tischler MB, Remple RK (2006) Aircraft and rotorcraft system identification: engineering methods with flight test examples. AIAA
18. Perfect P, Timson E, White MD, Padfield GD, Erdos R, Gubbels AW (2014) A rating scale for the subjective assessment of simulation fidelity. Aeronaut J 118(1206):953–974. https://doi.org/10.1017/S0001924000009635
19. Cooper GE, Harper RP (1969) The use of pilot ratings in the evaluation of aircraft handling qualities. NASA TN D-5153. NASA
20. White MD, Perfect P, Padfield GD, Gubbels AW, Berryman AC (2013) Acceptance testing and commissioning of a flight simulator for rotorcraft simulation fidelity research. Proc Inst Mech Eng Part G J Aerosp Eng 227(4):663–686. https://doi.org/10.1177/0954410012439816
21. Lu L, Padfield GD, White MD, Quaranta G, van't Hoff S, Podzus P (2023) Case studies to illustrate the rotorcraft certification by simulation process: CS 29/27 dynamic stability requirements. In: 49th European Rotorcraft Forum, Bückeburg, Germany

22. Hodge SJ, Perfect P, Padfield GD, White MD (2015) Optimising the yaw motion cues available from a short stroke hexapod motion platform. Aeronaut J 119(1211):1–21. https://doi.org/10.1017/S0001924000010228
23. Hodge SJ, Perfect P, Padfield GD, White MD (2015) Optimising the roll-sway motion cues available from a short stroke hexapod motion platform. Aeronaut J 119(1211):23–44. https://doi.org/10.1017/S000192400001023X

Chapter 13
Case Study 2: CS 29/27 Category A Confined Area Rejected Take-Off

13.1 Introduction

This Chapter presents the results from the case study on the Applicable Certification Rule (ACR) CS-27, Appendix C [Criteria for Category (CAT) A] and CS-29 paragraph 62 [(Rejected take-off (RTO): Category A] [1, 2], to illustrate the application of the guidance presented in this Book.

CS-27 Appendix C—Criteria for Category A, C27.2 states that:

> The following paragraphs of CS-29 must be met in addition to the requirements of this CS:
>
> 29.62 Rejected take-off: Category A which states that:
>
> The rejected take-off distance and procedures for each condition where take-off is approved will be established with:
>
> (a) The take-off path requirements of CS 29.59 and CS 29.60 being used up to the TDP (Take-off Decision Point) where the critical engine failure is recognised, and the rotorcraft landed and brought to a stop on the take-off surface;
> (b) The remaining engines operating within approved limits;
>
> ….

Further CS 29.51 Take-off data: General, states:

> that take-off data must be determined at each weight, altitude and temperature selected by the applicant

and for a CAT A Take-off, CS 29.53 states that:

> the performance must be determined so that if one engine fails at any time after the start of take-off, the rotorcraft can: …return to and stop safely on the take-off area.

For CAT A operations, the aircraft take-off weight is limited to a value such that if an engine failure occurs at, or before, the TDP the pilot will have to abort the take-off as the rotorcraft "*has not yet achieved sufficient energy to assure continued flight*" [2].

G.D. Padfield et al., *Rotorcraft Certification by Simulation and Analysis*, Springer Aerospace Technology, https://doi.org/10.1007/978-3-031-86398-1_13

Testing to determine the TDP starts with the aircraft first tested in a lightweight configuration which is then increased to determine the maximum take-off weight and c.g. configurations possible for the environmental conditions under consideration [2]. The procedure considers both the rejected take-off and the continued take-off, that may have different critical parameters. The flight trials require significant time and expense and pose safety risks; application of the Rotorcraft Certification by Simulation (RCbS) process can be used to reduce these factors.

In this Case Study, the use of simulation to achieve full credit for the CAT A RTO ACR for an AW109 Trekker is presented. This is an example of an 'ambitious' influence level, included to exercise the process at I4 with the expectations of only limited success or even failure. Section 13.2 briefly describes aspects of Phase 1 of the RCbS process for this ACR. Chapter 12.3 presents a flight-test-manoeuvre (FTM), in the style of an ADS-33E (ADS-33) mission-task-element (MTE) [3], that can be used for fidelity and certification assessments. Phase 2a of the RCbS process, Flight Simulation Model (FSM) build and development, is presented in Sect. 13.4. A description of the Flight Simulator (FS) build process is presented in Sect. 13.5 and some results from an exploratory piloted simulation trial are presented in Sect. 13.6. Section 13.7 then summarises the main conclusions and recommendations derived from this case study.

13.2 Phase 1: Requirements Capture and Build

Influence and Predictability. Phase 1 of the RCbS process contains subtasks for a selected ACR: selecting the appropriate Influence and Predictability (I-P) levels, defining the simulation types and critical features, and assembling their detailed requirements. The concepts of Influence, Predictability and Credibility levels convey the underlying consequences of the application of RCbS, in terms of safety and efficiency in the certification campaign, and they inform the FSM and FS requirements capture and build phases of the RCbS process.

In this Case Study, the applicant desires to only use simulation to determine the maximum take-off weight for the CAT A ACR, corresponding to I4 Full Credit, i.e. no flight tests will be conducted. The applicant wants to extrapolate from flight test data obtained at the low altitude (748 ft) Domain of Validation (DoV) condition to a high-altitude case (9280 ft). The Federal Aviation Administration's (FAA's) Advisory Circular (AC) 29-2C, § 29.45 (Amendment 29–24) Performance—General, 2(B) [4], allows for a maximum extrapolation of $\pm$ 2000 ft for "*height-velocity, and engine operating characteristics*". The extrapolation of > 8000 ft clearly puts this Case Study at the P3 predictability level, "*Extensive extrapolation in the Domain of Extrapolation (DoE)*".

13.3 FTM Development

A key element of the RCbS process is the use of FTMs for the assessment against the ACRs. FTMs can be developed with defined performance standards, and assessed by pilots using rating scales, including the Cooper-Harper handling qualities rating (HQR) scale [5]. An RTO FTM was developed in accordance with the procedures provided in the AW109 Trekker Rotorcraft Flight Manual (RFM) CAT A supplement [6]. The profile of the CAT A RTO manoeuvre in a *confined area* is shown in Fig. 13.1. The manoeuvre begins with the evaluation pilot (EP) starting in a 5 ft hover above a helipad and then initiating a rearwards and upwards climb (position 1 in Fig. 13.1) towards the TDP. Following an engine failure (position 2), the EP arrests the climb and rearwards motion, described as the 'bow', and the aircraft is flown in a controlled descent (position 3) and cushioned onto the helipad (position 4). Full details of the FTM are provided in Appendix 1. The FTM ground speed performance requirements were informed through discussions with an EASA test pilot and flight test engineer.

13.4 Phase 2a: FSM Build and Development

Flight test data and a FLIGHTLAB [7] FSM of the AW109 Trekker (F-AW109) were provided by Leonardo Helicopter Division (LHD) to exercise aspects of the RCbS process for a range of test conditions, including for validation purposes. Note that in the formal RCbS process, the flight test data would be gathered in Phase 2, in conjunction with the development of the FSM and FS, and following the development

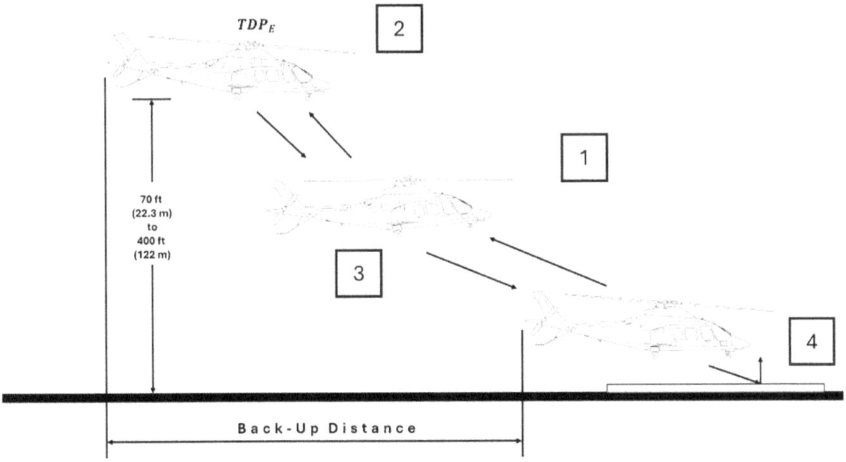

Fig. 13.1 CAT A RTO (Confined Area) profile, description in Ref. [6]

[including Verification & Validation (V&V)] of the Flight Test Measurement System (FTMS).

FSM Build. The FSM development phase follows a structured approach building up from the FSM component level to the aircraft level; the modelling complexity required depends on the application. For CAT A RTO simulations of conventional rotorcraft configurations, where performance is a major driver, it was considered that sufficient fidelity can be achieved using contemporary, state of the art, methods.

The baseline FSM had undergone some fidelity assessments by LHD for up-and-away flight conditions, but not for low-speed conditions in proximity to the ground. The FSM features a rigid, articulated, blade-element main rotor with nonlinear aerodynamics. The main rotor induced velocities are computed with a Peters-He three-state inflow model along with a source image ground effect model. The main rotor aerofoil data are available in the form of table lookups of the aerodynamic coefficients C_l, C_d and C_m, as functions of the angle of attack and Mach number. The blade airloads are computed in a quasi-unsteady fashion including unsteady circulatory effects from thin aerofoil theory. A Bailey rotor model is used for the tail rotor. Fuselage aerodynamic loads are computed at, and applied to, a single computational point. The fuselage and tail surface force and moment coefficients are available as functions of angles of attack and sideslip, derived from model-scale wind tunnel test data.

One of the objectives for the FSM development for this ACR was to obtain 'realistic' engine torque and rotorspeed responses within the constraints of the available power for the atmospheric conditions of the test day. Power limits were imposed based on available engine data and flight test data, aimed at determining the installation losses. Information on the engine control logic was not available, so the rotorspeed governor logic was emulated using a Proportional–Integral–Derivative (PID) controller. In practice, the unavailability of such data would likely preclude achieving I4 influence. The controller gains were originally tuned to up-and-away flight test frequency response data but were modified for the One-Engine-Inoperative (OEI) scenario.

The control laws of the Stability Augmentation System and Attitude Hold modes were replicated using a conventional PID control architecture as provided by LHD.

The skid landing gear was modelled as a set of four spring-damper struts with default FSM FLIGHTLAB ground contact friction parameters. As the intent of the simulation was not to determine landing gear loads during touchdown, only the entry conditions at the instant of ground contact were considered. It is acknowledged that, in practice, determining the varying loads on the undercarriage during touchdown would be included in the certification.

FSM Validation. A dedicated validation effort is needed for the application of the FSM to CAT A RTO simulation. Elements of such validation may include:

- Performance in trim:
  - Hover IGE
  - Low-speed backward climb

  - Steady OEI descent (at target main rotorspeed, N_R).
- Control response:
  - Hover and low-speed steps/doublets on all axes
  - Hover and low-speed frequency response data (e.g., from control sweeps and system identification).
- Engine and RRPM response characterisation:
  - Torque and N_R recovery after single engine failure (ground or flight test).
  - Flare effectiveness flight tests.

Some of these data were available in the set provided to the RoCS project (e.g. trim) that enabled limited validation. A key question regarding the fidelity of the FSM is what metric should be used for validation purposes. The requirements detailed in Certification Specification for Flight Simulator Training Devices (Helicopter), CS-FSTD(H) [8], were taken as a 'candidate' fidelity metric under the justification that the standards are suitable from a piloting perspective. Other metrics are available and their use would need to be justified to the certification authority.

Figure 13.2 shows a comparison of flight test (FLT) and FSM trim data in the hover and low speed regime. The error bars for the flight test data represent the standard deviation (a component of u_r) of the parameter observed over a 10 s trim time history. As can be seen, the baseline FSM exceeds the CS-FSTD(H) tolerances in different fore/aft flight regimes and warranted further development to improve its fidelity.

The aerodynamic interaction between the main rotor wake and the fuselage remains one of the more challenging phenomena to accurately model in low-speed forward and rearward flight conditions. Although a model can be 'tuned' to achieve an 'acceptable' match between flight test and simulation data in terms of power and thrust required, the RCbS framework strives for a more physics-based modelling approach. The fuselage interference modelling in the FLIGHTLAB FSM [7] uses a lookup table that provides the aerodynamic force coefficients as functions of the angles of attack and sideslip. In the baseline FSM, a single Aerodynamic Computational Point (ACP) was used, disregarding the distribution of interference velocities and cross-sectional area of the fuselage. To account for these effects empirically, multiple ACPs were defined by a set of locations along the length of the fuselage. An improved match between FLT and the updated FSM for pitch attitude (Fig. 13.2a) and engine torque (Fig. 13.2d) was obtained using 16 equally distributed ACPs with a weighting based on segmented fuselage volumes, as shown in Fig. 13.3. The outstanding mismatch in collective lever (approx. 10%) and pitch angle would need to be resolved and corrected in a 'real' case, but, for the demonstration of the RCbS process, the FSM was judged to have sufficient fidelity for use in the next phases.

The updated FSM was used to assess the dynamic response fidelity when 'flying' the RTO FTM. To facilitate the predicted simulation fidelity assessment, a virtual pilot model was used [9]. The pilot model enables the definition of the piloting strategy

Fig. 13.2 Impact of FSM updates on correlation against flight test for low-speed Out of Ground Effect (OGE) level flight trim [shaded areas indicate CS-FSTD(H) tolerances for low-speed handling qualities]

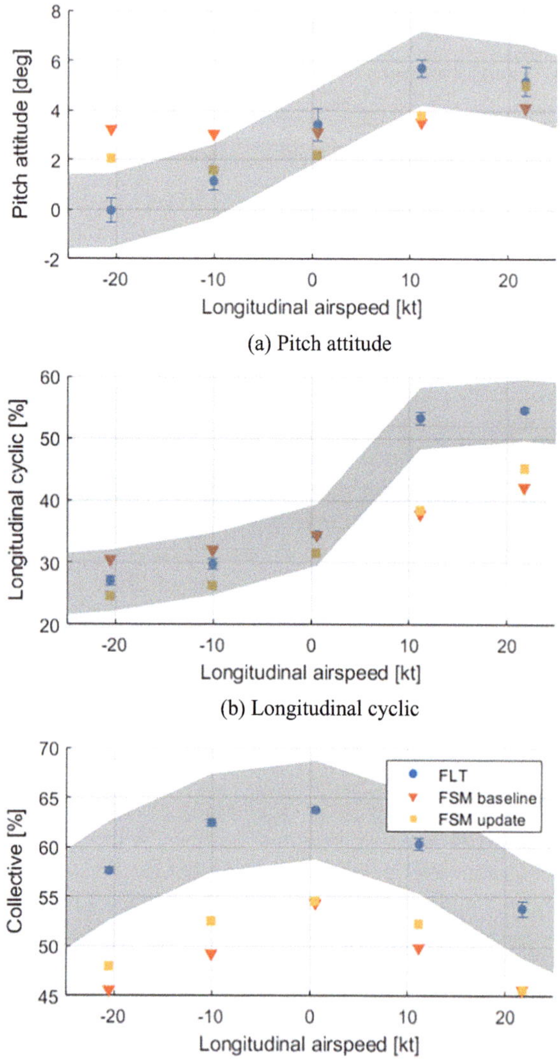

(a) Pitch attitude

(b) Longitudinal cyclic

(c) Collective levera

e.g., power management, inceptor activity, to support validation against flight test data. The parameters pertinent to the different phases of the RTO FTM are:

- Hover:
  - Height above take-off surface
- Climb to TDP:
  - Vertical speed

Fig. 13.2 (continued)

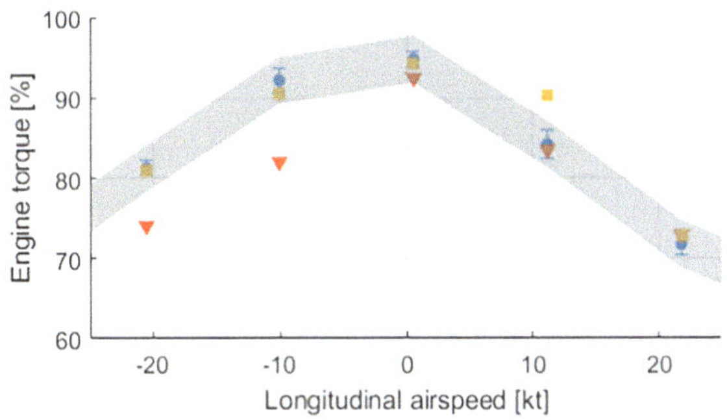

(d) Engine torque

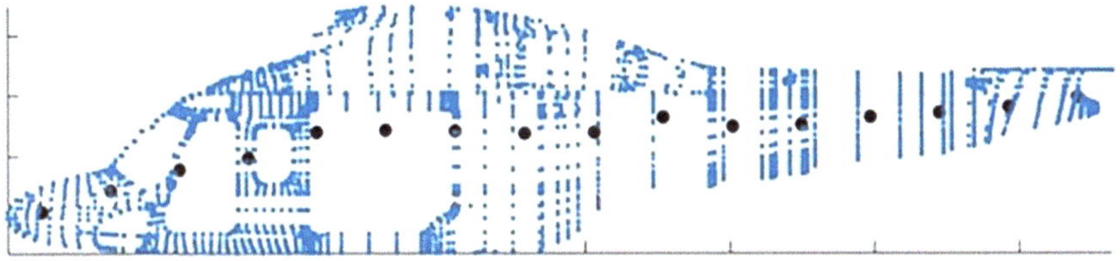

Fig. 13.3 ACP distribution (black markers) used for rotor-fuselage interference computations

  - Climb gradient
  - Engine failure height

- OEI descent to helipad: Pilot intervention delay

  - Target OEI main rotorspeed N_R or delta-collective

- Landing flare:

  - Collective flare lead time
  - Target touchdown rate of descent
  - Target touchdown ground speed.

During the OEI descent, the collective can be used to adjust main rotorspeed, N_R, in a closed-loop manner, or be held constant at a certain offset below the climb-out value in accordance with AW109 RFM procedure. The virtual pilot control logic features nested PID controllers on each of the four aircraft axes. Triggered and driven by the parameters listed above, the heave/collective axis is controlled based on vertical speed or OEI rotorspeed, whereas the pitch/cyclic axes are governed by ground speed through the target climb/descent gradient.

Figures 13.4, 13.5 and 13.6 show the validation results of the FSM against flight test data for one of the low-altitude flights available within the DoV. Again, the grey shaded areas indicate the 'fidelity' tolerances defined in CS-FSTD(H). In this example, the simulated trajectory is initiated at the start of the stabilised climb. The engine failure is timed to coincide with flight test (represented by a blue circle in the figures) and the virtual pilot reacts to the failure after a specified intervention delay by slightly lowering the collective, if needed, to achieve an OEI N_R of 101% in accordance with the RFM procedure. The engine/rotorspeed governor reacts by demanding full power of the operative engine (aiming to achieve 102% N_R). At a defined lead time prior to projected touchdown, the skids are levelled, the rate of descent is arrested and the rotorspeed decays. The simulation is halted when weight on skids is detected, prior to the rotorspeed recovery observed in the flight test data. Note that the touchdown (at the front edge of the helipad) is performed with residual forward ground speed and with the skids approximately level.

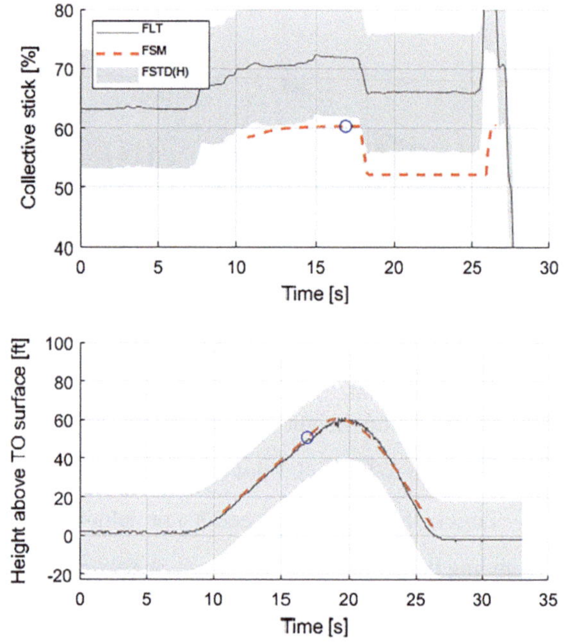

Fig. 13.4 Correlation between simulation and flight test for CAT A RTO trajectory at low-altitude test site: collective lever and height [shaded areas indicate CS-FSTD(H) tolerances for low-speed handling qualities]

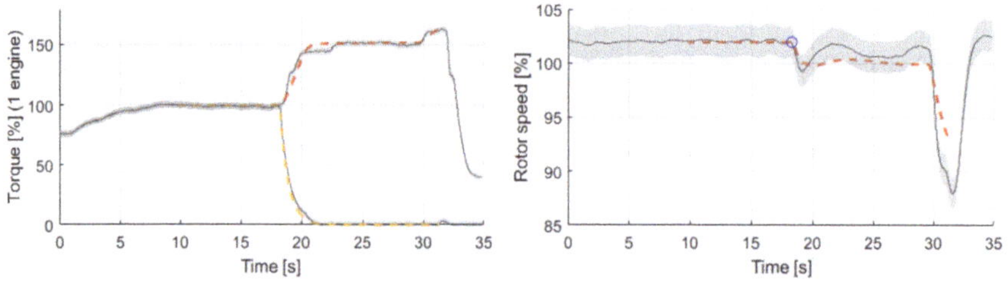

Fig. 13.5 Correlation between simulation and flight test for CAT A RTO trajectory at low-altitude test site: engine torque and rotorspeed [shaded areas indicate CS-FSTD(H) tolerances for low-speed handling qualities]

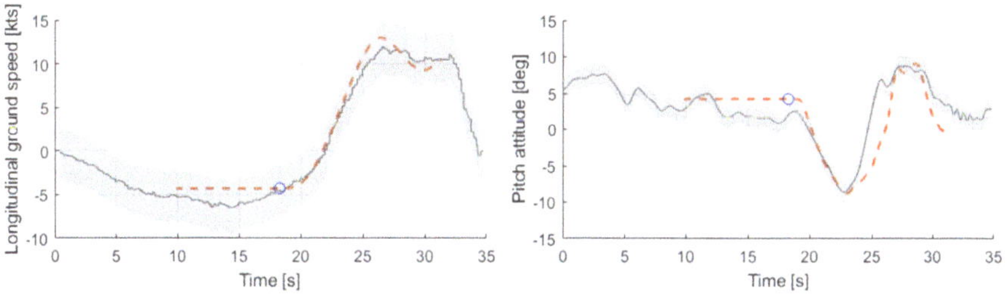

Fig. 13.6 Correlation between simulation and flight test for CAT A RTO trajectory at low-altitude test site: ground speed and pitch attitude [shaded areas indicate CS-FSTD(H) tolerances for low-speed handling qualities]

Several observations can be made about the correlation achieved. Firstly, there is a notable offset in the collective trace, consistent with the trim results presented in Fig. 13.2c. Given that the collective control limits are not a driving factor for this ACR, the discrepancy can be accepted, although, as noted above, would need further investigation in a 'real' application of RCbS.

There are also notable differences in the pitch attitude between the flight test and simulation, starting from trim at approximately 12 s. The correlation is highly sensitive to the centre of gravity position which has not been corrected for fuel expenditure in the simulation. The pitch attitude does not significantly affect aircraft performance but does influence the visibility of the take-off and landing surface through the chin window, which is a crucial factor in the ability of the pilot to perform the RTO procedure. Hence, piloted assessment would be required to confirm that the attitude prediction is suitably representative of the real aircraft.

A further observation can be made about the OEI N_R response. The prediction remains within the CS-FSTD(H) tolerance, although the flight test shows more transient changes between 25 and 30 s. The FSM generally shows a less dynamic N_R response to the change in attitude and, hence, tip-path-plane tilt. Such behaviour warrants more detailed investigation involving, e.g., comparison with in-flight flare effectiveness test data. Finally, it is worth reiterating that the CAT A RTO data available in RoCS were not historically gathered for the purpose of FSM fidelity assessment and validation. Missing information that would have been useful for validation purposes includes control rigging measurements, accurate position data to compare 3-D trajectories, wind mast measurements to confirm the requisite calm conditions and rotor flapping for tip-path-plane correlation.

13.5 Phase 2b: FS Build and Development

A pilot-centred approach is adopted in the RCbS process (Phase 2b) wherein the EP is provided with 'sufficient' cues, from the FS features, to complete the task. This should enable the EP to achieve the same level of task performance as in flight with minimal control strategy adaptation. This approach draws upon the Simulation Fidelity Rating (SFR) scale methodology developed in Ref. [10].

The FS build in Phase 2b requires inputs from Phase 1 in terms of the ACR and IP levels and includes 'Engineering Design Data' in the form of, but not limited to:

- flight control mechanical characteristics
- instrument panel displays
- warning lights/sounds and aural cues.

Input is also required from the FTMS and DoV flight test activity to ensure that procedures used, and data gathered, in flight are replicated in the simulator for validation or compliance testing.

Workup trials conducted using the HELIFLIGHT-R flight simulator [11] at the University of Liverpool (UoL), highlighted the following important FS features to cue the EP in the CAT A RTO FTM:

- the Visual Motion Cueing System (VzMCS) i.e., the outside world visual scene content,
- the Vestibular Motion Cueing System (VeMCS)—note that both the VzMCS and VeMCS provide the EP with 'motion cues' that need to be consistent,
- the crew station layout and structure, in particular the instrument panel and the external visibility characteristics of the crew station (e.g. chin window)
- the sound cueing system e.g., main rotor noise, low and high rotorspeed audio alerts, annunciation of failures,
- the flight control inceptors and related forces.

Another consideration in the RCbS process is the role and experience of the EPs who aid in the RTO FTM and FS development. As acknowledged in the SFR

Fig. 13.7 Post engine failure eye tracking during the descent phase of the FTM

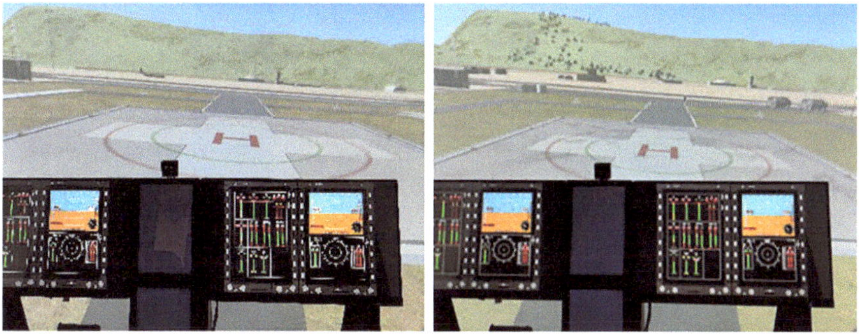

(a) (b)

Fig. 13.8 Original (**a**) and enhanced (**b**) VzMCS for the start of the RTO FTM

assessment process, the EPs need to have real-world experience with the FTM, and on the test aircraft, to provide confidence in the FS development and assessment activities.

VzMCS Development. The VzMCS feature provides the EP with visual motion cues. Definition of the VzMCS feature sufficiency requirements can be considered under several categories e.g., Field of View (FoV), micro- macro-textures (e.g. pixels per arc/s), and image generation transport delay. CS-29 sets rather general rules on this topic, e.g. paragraph 773, "*Pilot compartment view*" states that:

> Each pilot compartment must be arranged to give the pilots a sufficiently extensive, clear, and undistorted view for safe operation.

Some further guidance on the "*Pilot compartment view*" requirement is provided in AC-29 paragraph 773 [4] which states:

> ....
>
> (v) For steep rejected take-offs and steep approaches such as used for oil rigs or confined heliports, the visibility should be such that the pilot can see the touchdown pad and sufficient additional area to the side and forward to provide both an accurate approach to the touchdown point as well as a satisfactory degree of depth perception.

Two aspects of the VzMCS were examined for this ACR: FoV and the degree of depth perception provided by micro- and macro-textures, acknowledging that for the latter there is no objective requirement defined.

Testing at the DLR simulation facility [12], the Level C FoV requirement of 150° × 40° horizontal/vertical FoV [8] did not provide sufficient vertical FoV for maintaining sight of the helipad during the manoeuvre. A horizontal FoV of less than 180° has the consequence of restricting peripheral cues for the EP. In testing at the DLR and UoL simulators, with FoV's 240° × 93° and 230° × 70°respectively, the inclusion of chin windows provided the additional vertical FoV that the EP required to maintain full sight of the helipad following the engine failure. The ability of the EP to maintain the helipad in the FoV was assessed at UoL using eye tracking that allowed real-time and post-sortie analysis of the EP's scanning pattern during the test. Figure 13.7 shows a still image from a video recording of an EP's scan following the engine failure and descent to the helipad. The green dot shows the pilot's gaze point, indicating that he was able to maintain the helipad in view during this phase of the descent.

Regarding micro- and macro-texture (VzMCS) fidelity, two EPs flew workup trials at UoL, without the VeMCS feature active. The EPs reported that the ground texture on the helipad and in the near and far field (Fig. 13.8a) did not provide sufficient detail to aid precise flightpath control in the final phases of the FTM, the flare and landing. Following their assessment, the helipad texture detail was enhanced (Fig. 13.8b) with additional texture layers to increase the scene content and additional medium and far field features e.g. vehicles, trees on distant hills to add depth cues. The EPs reported that the additional scene content improved their 'perception' of height and vertical motion in the final phase of the FTM and hence their ability to achieve improved task performance.

VeMCS Development. A key element of the VeMCS feature is the design of the Motion Drive Algorithms (MDA). Previous research at UoL has shown that careful selection of the MDA parameters i.e., gains (k) and high pass break frequencies (ω_{hp}), is essential to provide both sufficient and 'correct', i.e. not adverse, VeMCS cues [13–15]. The research has shown that sufficient vestibular motion cues could be achieved by careful harmonisation of the motion filter gains for yaw-sway [13], roll-sway [14] and pitch-surge tasks [15]. Reference [16] proposes a framework for the systematic tuning of motion filter parameters for a simulator flying task. A similar 'tuning' approach was adopted in the current Case Study considering both the primary and secondary axes of the FTM. The pitch-surge motion filters were initially tuned from a default set before further tuning was conducted to provide additional yaw and heave cueing following the engine failure and in the final flare (Appendix 2).

Subjective assessment of the VeMCS feature was undertaken using the Motion Fidelity Rating (MFR) scale [13] in advance of the RoCS testing. A 'default' MDA tuning set which had been derived for general rotorcraft flying i.e., not for this FTM, and an MDA set tuned for the RTO FTM were assessed by two EPs against a case with the VeMCS disabled.

Both pilots reported that enabling the VeMCS feature improved task performance in the RTO FTM, allowing for improved control of ground speed and rate of descent when approaching touchdown. However, the 'default' tuning set gave adverse cues to the pilot in the bow and flare phases of the FTM where the largest and most rapid pitch changes were used.

Both pilots also cited a lack of yaw and heave cues after the engine failure as deficiencies with the default MDA tuning. These issues were addressed in the RTO-tuned set by increasing heave gain and break frequency, increasing the yaw gain, and reducing the pitch break frequency. The RTO-tuned MDA parameters allowed improved task performance e.g. touchdown position and velocity, and improved the perceived fidelity of the VeMCS feature. Both EPs commented on poor heave cueing in the final phase of the FTM, attributed to the disengaging the motion platform during the final flare to prevent any undue structural loading to the projectors; a point that was revisited in the formal RoCS trial.

Additional FS Development. The AW109 Trekker uses the Genesys Aerosystems IDU-680 instrument panel [17]. A replica of this panel was developed and implemented in the HELIFLIGHT-R simulator using Presagis' VAPS XT software (Fig. 13.9). An initial implementation was developed based on information from the publicly available IDU-680 user manual [17] which describes a generic version of the IDU-680, not one specific to the Trekker. The panel was updated during workup testing based on EP feedback e.g., removal of 'clutter' around the attitude indicator to provide clearer indication of pitch information.

An important element of the RTO testing was to ensure that a 'valid' test condition was evaluated. This is based on the flightpath profile i.e., a climb rate of approximately 350 ft/min and a rearwards ground speed of 4–5 kts prior to the engine failure (see Appendix 1). Flightpath profiles outside these values can be considered as test points for so-called 'abuse case' testing, to explore the impact of variations in the standard flight manual method, but the aim of this testing was to examine certification for a 'standard' flightpath profile. To assist with the assessment of this 'standard' profile, a performance display was added to the Genesys multi-function display (Fig. 13.10). The FS operator could use the performance display during flight to provide real-time flightpath information e.g., call-out of the EP being above/below the desired flightpath (lower right indication in Fig. 13.10). It is noted that, in current certification practice, flight test engineers use a similar display to help in two ways, (1) to assess if the manoeuvre is being flown correctly and (2) to train the pilot to stay on the trajectory by using the external cues (e.g. what s/he sees through the chin window). However, following familiarisation, the pilot is expected to fly the manoeuvre without the help of the test engineer. As shown in Fig. 3.10, the current value of a performance variable is indicated by a blue cross in each of the displays. During flight, the EP could not see this panel, but could access it after a test point to judge their performance at TD, shown by a white 'X' on the display.

Another element that was developed during the workup trials was an automatic engine failure engagement which could be triggered at a defined height above the ground. This was developed to provide consistency in the engine failure point during testing. It is acknowledged that in real-world testing there would be variations in the failure height, and this could be included in future 'abuse case' testing.

Based on EP feedback in workup testing, it was noted that variations in N_R of $\pm$ 1–2% could be detected through audio cueing. HELIFLIGHT-R's Audio Cueing System feature, SimAudio, was improved to modulate the audio file of the rotor noise to better cue N_R variations. SimAudio was also upgraded to enable failure aural cues to be provided directly to the EP via their headset to reduce any potential delay in recognising and responding to an engine failure.

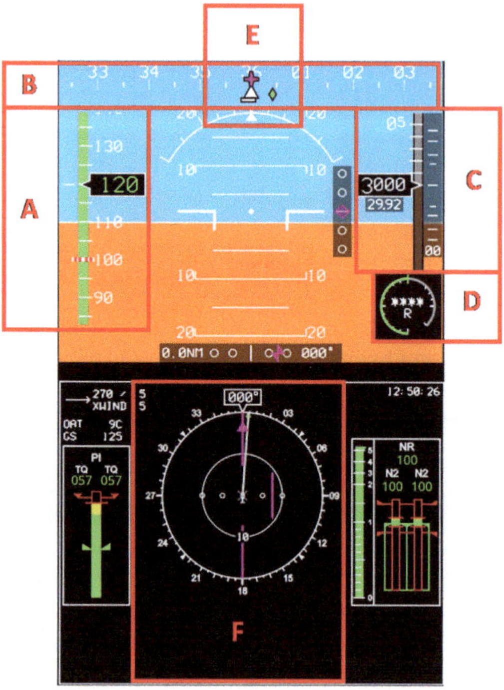

A – Airspeed Display

B – Heading Display

C – Altimeter and Vertical Speed Indicator

D – Radar Altimeter

E – Slip/Skid Indicator

F – Horizontal Situation Indicator with Lateral Track Pointer (Magenta)

Fig. 13.9 Genesys Instrument Panel generated using Genesys [17]

13.6 Piloted Simulation Assessment

The aim of the simulator testing presented in this Case Study was to illustrate how an applicant might achieve the I4-P3 approval for a CAT A RTO. As part of the Phase 2b V&V, a fidelity evaluation of the FS was undertaken using the SFR scale [10]. The SFR scale allows the EP to assess the FS fidelity based on a comparison of the task performance achieved in flight and in the FS, together with an assessment of any adaptation of control strategy and compensation. Clearly, (recent) experience with the real aircraft, and the type of testing, in flight is a pre-requisite here. To aid in the FS fidelity assessment, HQRs and SFRs were awarded by the EP, supported

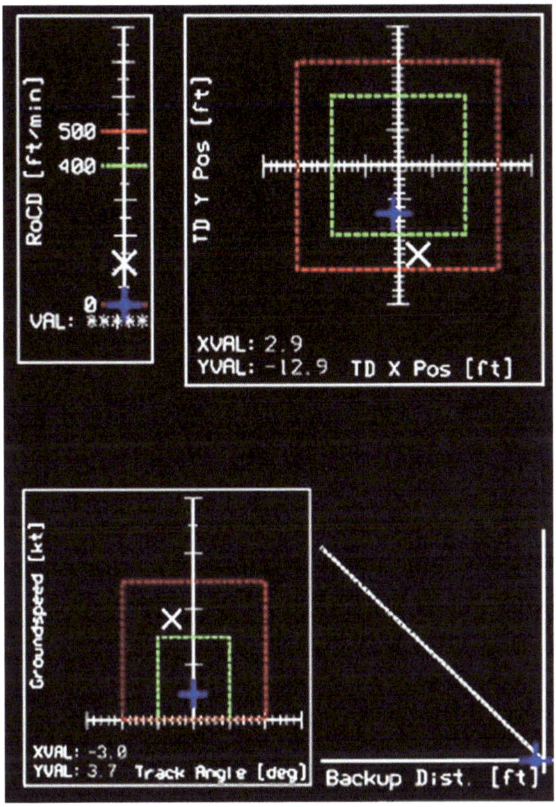

Fig. 13.10 RTO Performance Parameter display

by comments obtained using an in-cockpit questionnaire (ICQ) (Fig. 11.7); it should be noted that 'equivalent' HQRs had not been awarded in flight test, but one of the EPs considered that they were sufficiently familiar with the RTO manoeuvre in the AW109 Trekker that an SFR was given.

The ICQ has been designed to elicit structured feedback on:

- FS feature fidelity e.g., visual and vestibular cues,
- HQR metrics e.g., task performance and pilot compensation,
- the phase of the FTM 'driving' the HQR,
- the award of other ratings, e.g., Pilot Induced Oscillation ratings and MFRs.

The EP's assessment of the task performance achieved was supported by feedback from the operator using the display shown in Fig. 13.10, and by a non-handling pilot in the FS left-hand seat. The role of the non-handling pilot was defined prior to testing as:

- Any duties as directed by the EP,
- Note task performance and be prepared to provide additional feedback to the EP if required,

- Monitor control inputs for frequency and amplitude and provide feedback if required,
- Monitor cockpit motion, visual and aircraft flightpath anomalies and feedback if necessary,
- Ensure copies of the FTM description and relevant rating scales and questionnaires are in the cockpit,
- Provide feedback, if required, during the relevant item in the ICQ.

DoV Testing. Testing was conducted at an aircraft mass of 3115 kg at a pressure altitude of 748 ft. A comparison of the flight and FS data is presented in Fig. 13.11 and Fig. 13.12. The dashed vertical line indicates when the engine failure occurs, and the grey shaded area represents the CS-STD(H) fidelity tolerances for comparisons with the flight test data.

In the RCbS process, comparisons between flight test and FS data would be made during the Phase 2b V&V and fidelity assessment activities. As shown in Sect. 13.4, the FSM-FT matches displayed areas of both sufficient and insufficient fidelity. Differences in control activity, response and performance observed between FT and FS, e.g. as shown in Figs. 13.11 and 13.12, could stem from either the FSM mismatch or an FS fidelity issue. It should be noted here that the EP who conducted the FS testing was not the same EP who conducted the flight test. Furthermore, the flight test was not conducted using the RCbS-developed FTM, but rather flown solely using the RTO procedure described in the RFM. These factors complicate the fidelity assessment and, while no firm conclusions can be drawn on the relative impact of FSM and FS degraded fidelity, some general observations can be made.

Differences between flight and simulation are evident in the control strategies used following the engine failure. In both cases, the pilot makes a forward longitudinal stick input and lowers the collective following the engine failure, shown in Figs. 13.11a and 13.12a, but the order in which the control inputs are applied are reversed in flight and in simulation. This difference in control strategy then affects the aircraft response shown in the pitch attitude (Fig. 13.11b), ground speed (Fig. 13.11c) and RoC/D (Fig. 13.12b).

The results show that rates of descent prior to TD exceed 800 ft/min in both flight and simulator. The F-AW109 FSM 'raises a flag' when the conditions for vortex-ring-state (VRS) are entered (RoD $\geq$ 800 ft/min in present case), although the consequent loss of vertical control is not represented correctly in the FLIGHTLAB version used in Liverpool. The RoCS team consider that the risk of entering VRS in this FTM needs to be given much more attention and the FSM developed to more accurately represent behaviour approaching and following entry.

Figure 13.13a shows that engine torque responses in flight and simulation are similar until the final phase of the manoeuvre. In both cases, the EP achieves the requirement of maintaining 101% N_R following the failure until the last few moments. The differences between flight and simulation N_R time-histories are a consequence of the different control strategies discussed above.

In the absence of a thorough 'u_r' analysis, some of the FT data are questionable, e.g. at around 34 s, while groundspeed has reduced to zero, the RoD is about 100 ft/

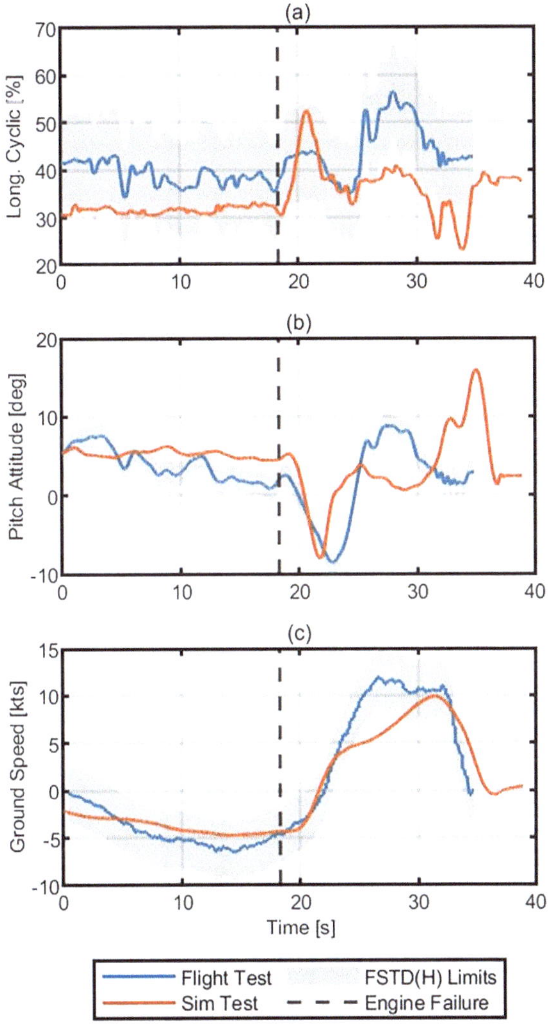

Fig. 13.11 Comparison of flight and simulator test data for (**a**) longitudinal cyclic, (**b**) pitch attitude and (**c**) ground speed [shaded areas indicate CS-FSTD(H) tolerances for low-speed handling qualities]

min. In FT, the pilot appears to have stopped moving the controls around 32 s. These kinds of anomaly will need to be resolved in the V&V of the FTMS in Phase 2.

A subjective fidelity assessment of the FS, with VeMCS-engaged, was conducted using the SFR scale (Chap. 11 and Ref. [10]) at an aircraft mass of 3040 kg. The EP awarded an SFR 5, indicating that "*similar performance*" was attainable with a "*moderate adaptation of task strategy*". Similar task performance relates to plan-position, vertical and ground speed at TD. In terms of task strategy adaptation, the EP reported that he was conducting the FTM in a "*similar manner*"' to that used in flight for phases 1–3 in Fig. 13.1. However, during the final flare phase (4 in Fig. 13.1), there was a lack of "*ground rush*" motion cueing which led to a moderate adaptation with the EP "*looking for other cues*" to aid control in this phase of the FTM. As noted

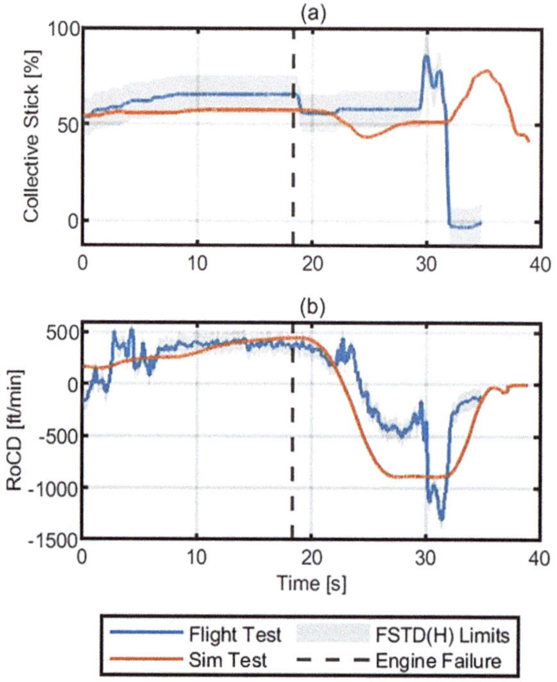

Fig. 13.12 Comparison of flight and simulator test data for (**a**) collective, (**b**) rate of climb/descent (RoC/D) [shaded areas indicate CS-FSTD(H) tolerances for low-speed handling qualities]

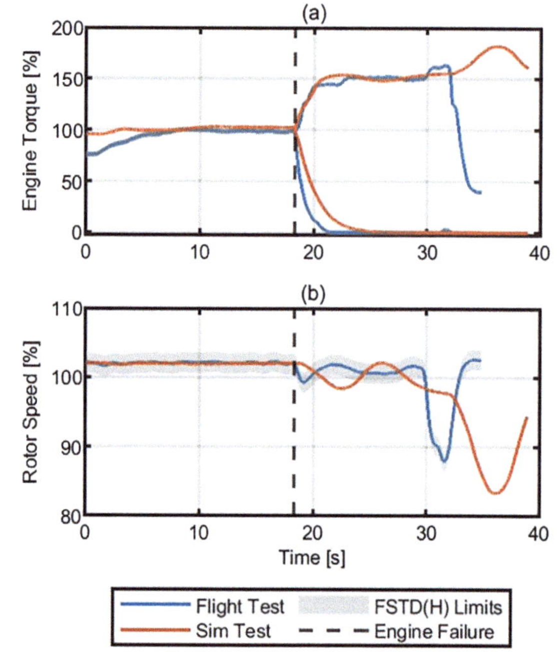

Fig. 13.13 Comparison of flight and simulator test data for (**a**) engine torque, (**b**) main rotorspeed, N_R [shaded areas indicate CS-FSTD(H) tolerances for low-speed handling qualities]

above, in the HELIFLIGHT-R testing, it has been standard practice to disengage the motion base during the final moments of the FTM to prevent any potential projector misalignment issues following 'hard' contact with the ground. The EP was asked to rate the FS fidelity, ignoring the final flight phase (position 4 in Fig. 13.1), and awarded an SFR 2, i.e. cueing fidelity was 'sufficient' for this FTM for the proposed certification I-P level.

The simulator results demonstrate that similar RTO profiles were achieved compared with flight test, but comparison is difficult when different piloting strategies have clearly been applied, and the tasks have not been flown to 'equivalent' performance standards, or by the same pilots. Furthermore, no HQR was available for the flight test case, as it was not flown to defined FTM performance standards. These factors make objective assessment of FS fidelity difficult, but subjective metrics such as the SFR are still useful tools for the FS fidelity assessment.

The utility of the VeMCS was investigated through a comparison of testing with motion-off and motion-on (using the RTO tuned MDA values) in the DoV, as shown in Figs. 13.14 and 13.15. The figures show results for multiple practice runs (denoted by the coloured x's) and the rated run (indicated by the squares). In the motion-off case, the EP required a larger number of runs prior to the award of an HQR compared with the motion-on case. A larger scatter of TD parameters is also noted (Fig. 13.14b), the increased range of which is illustrated by the upper and lower black bars in Fig. 13.15 compared with the motion-on case testing.

Whilst an HQR 4 was awarded in both the motion-on and -off cases, the data show marginal desired performance was achieved with motion off. The EP reported that he could better control flightpath with the motion-on and the TD RoC/D and ground speed were well within desired performance, compared with the motion-off case, where borderline desired performance was achieved. The performance achieved was confirmed by the non-handling pilot. An MFR 5 was awarded for the motion-on case indicating that "*useful*" VeMCS cues were provided for the task. In both cases, the EP reported that 101% N_R could be maintained following the failure and that the driving phase of the FTM for the HQR was the flare to TD.

Despite the uncertainties stemming from the fidelity assessments, the FS results provided sufficient confidence to conduct testing in the Domain of Extrapolation (DoE).

DoE Testing. The section above has focussed on the fidelity assessment of the FS which is a critical part of the RCbS process, but the ultimate goal of this Case Study was to attempt to achieve full credit in the DoE (I4-P3); density altitude increased from 748 ft (DoV) to 9280 ft (DoE).

The DoE tests were conducted with the motion-on throughout the FTM. An incremental approach was used to determine the maximum take-off mass for the CAT A rejected take-off in the DoE. Starting at a mass of 2430 kg, a minimum of three test points were flown before the mass was increased, in 25 kg increments, up to a maximum value that could be achieved. Figure 13.16 shows the range of the longitudinal (Y) position, ground speed and RoC/D at TD. Figure 13.7 shows the maximum

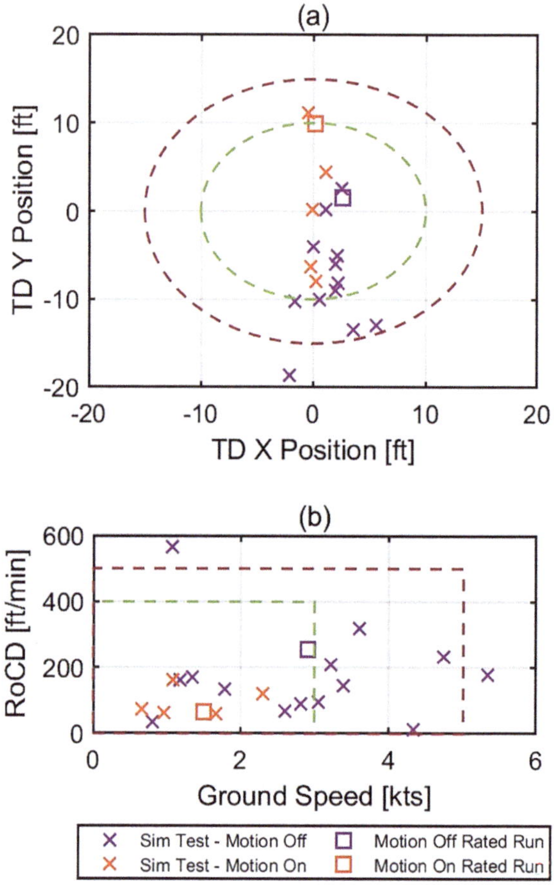

Fig. 13.14 Effect of motion off/on for (**a**) TD positions and (**b**) RoC/D during DoV testing (green dashed line: desired, red dashed line: adequate)

engine torque used after the engine failure. The test point scatter is indicated by the upper and lower limits of the bars.

At a mass of 2430 kg, the EP noted he was within desired performance for all parameters, achieved 101% N_R following the failure, with moderate control compensation in the longitudinal cyclic during the initial bow, and the final flare of the FTM. As the aircraft mass was increased to 2480 kg, there was a small change in the range of the longitudinal TD parameters, all still within desired performance values. An increase in the maximum torque used following the engine failure can be observed. Increasing the mass to 2505 kg resulted in an increase in the range of longitudinal touchdown positions, Y (Fig. 13.16a) and the maximum torque used has increased (Fig. 13.17), with the EP reporting a noticeable reduction in engine performance margin.

Increasing the mass to 2530 kg (Fig. 13.18), the EP stated that whilst longitudinal position performance was "*borderline*" desired, he was encountering issues with maintaining N_R whilst applying collective in the flare at TD. The EP's assessment

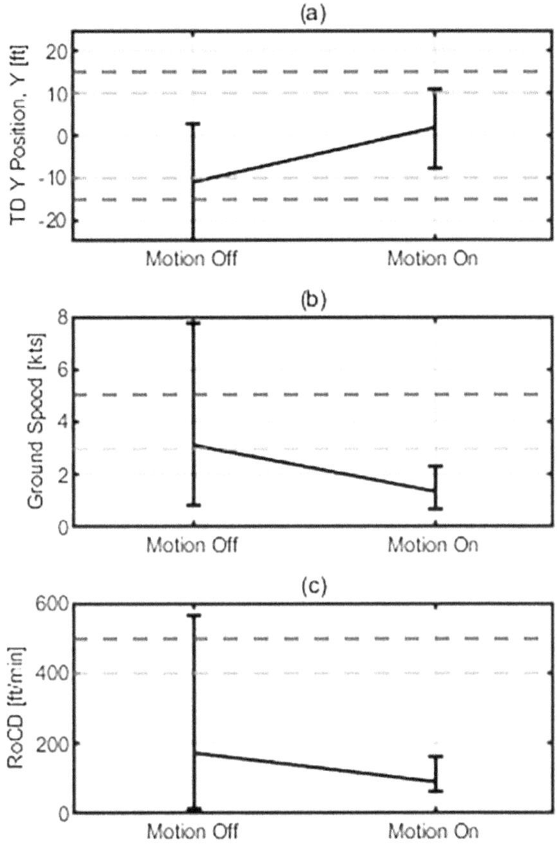

Fig. 13.15 Effect of motion off/on scatter for (**a**) TD position, (**b**) ground speed and (**c**) RoC/D during DoV testing (green dashed line: desired, red dashed line: adequate)

of task performance was confirmed by the non-handling pilot. In the EP's opinion, the aircraft at this mass would not be certified for this ACR. The torque excursions beyond maximum transient (Fig. 13.17) would also be unacceptable in certification.

The outcome of the testing was that the EP stated that "*full credit*", without considering DoV predicted fidelity deficiencies, could be expected for this ACR in the DoE at a mass of 2505 kg, at the extrapolated high-altitude condition, using the FS environment developed in this Case Study.

This is very much a preliminary result, as procedure variability testing would normally be required for 'clearing' this mass for certification based on, for example, variations in climb rate in the climb out phase, and engine failure height. Nonetheless, from communication with LHD, the mass identified as the limiting case for the DoE extrapolation is similar to that certified for the aircraft, as documented in the RFM's weight-altitude-temperature charts.

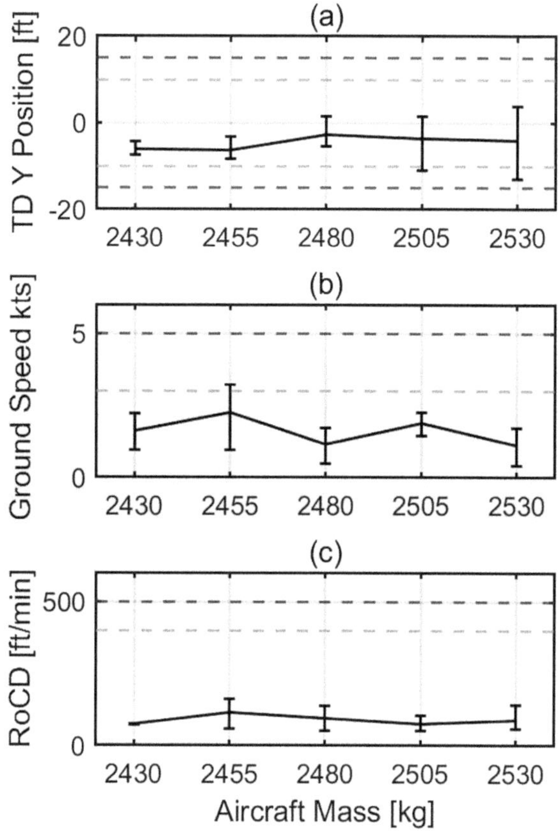

Fig. 13.16 Range of Y position (**a**), ground speed (**b**) and RoC/D (**c**) after engine failure with increasing aircraft mass during DoE testing (green dashed line: desired, red dashed line: adequate)

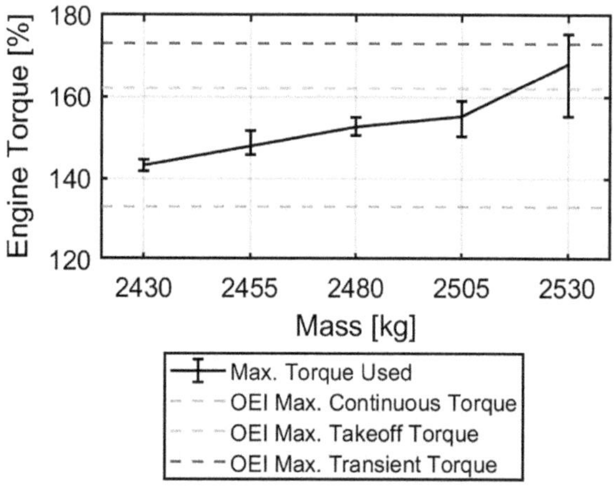

Fig. 13.17 Range of maximum single engine torque used after engine failure with increasing aircraft mass during DoE testing

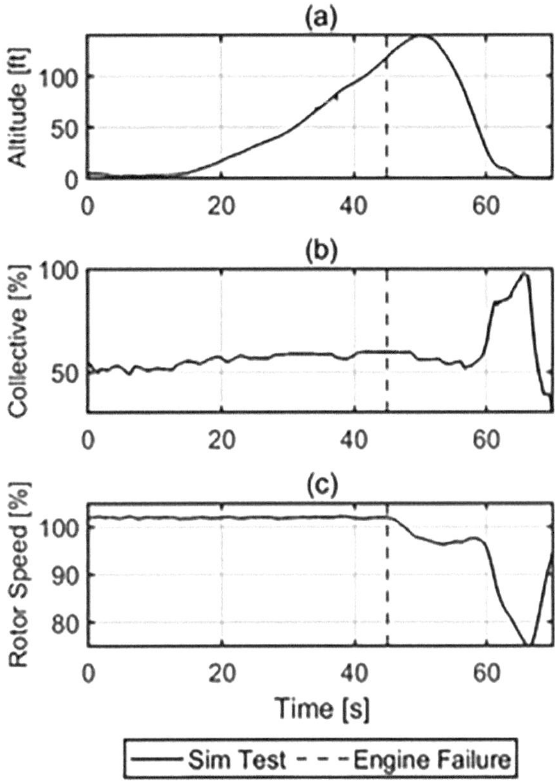

Fig. 13.18 Time histories of altitude (**a**), collective (**b**) and main rotorspeed, N_R, (**c**) for the 2530 kg test condition

13.7 Concluding Remarks and Recommendations from the RTO Case Study

This Chapter has reported on exercising elements of the RCbS process for the confined-area Category A Rejected Take-off ACR, as expressed in the EASA certification specifications CS27 and CS29. A general conclusion is that, despite uncertainties encountered in the fidelity assessment, the FTM developed in this case study demonstrated the significant value of RCbS for this ACR. The exercise can be considered an extraction of the fidelity assessment in Phase 2b, highlighting fidelity deficiencies that would need 'fixing' prior to Phase 3. The DoE exercise to establish the maximum weight at high altitude received a full credit result from the EP, with the noted caveats.

Detailed conclusions from the Case Study are as follows.

(i) An FTM has been designed to enable Phase 2b FS fidelity assessment to be conducted. As with all the FTMs developed in this Book, further consideration should be given as to how the FTM might capture those elements derived from normal certification practice for different aircraft.

(ii) The use of a virtual pilot model to aid in the tuning of the FSM is a viable step in the V&V process for this ACR, highlighting the potential of such methods for strengthening the predicted fidelity analysis.
(iii) The VeMCS feature, and the MDA configuration used, had a positive impact for this ACR in terms of control of touch-down speed and plan position.
(iv) The SFR scale provides a useful tool for assessing the (perceived) fidelity of the FS for the chosen ACR.
(v) Extrapolation from the DoV to DoE conditions, in terms of altitude/weight increase, has demonstrated the feasibility of the RCbS process applied to the RTO ACR.

The results in the Case Study, whilst promising, pose additional questions that provide 'fruitful' areas for exploration. For example, what would be an acceptable HQR for an ACR, if adopted in the certification process? Another area for investigation concerns how MDAs might be designed to match with the dynamics of an FTM/ACR. These kinds of question can be addressed in the continuing development of the RCbS process and its application by early adopters.

Appendix 1: RTO FTM Description

Mission	Civil Transport			
Critical HQs	Vertical velocity and N_R response to collective, pitch/roll response to cyclic; cross-couplings: pitch/roll, roll/pitch, collective/yaw, collective/pitch			
Objectives	• Check ability to perform steady climb to Take-off Decision Point • Check ability to return to a helipad after failure of one engine, while controlling vertical descent rate and forward speed with longitudinal cyclic and lateral track and roll angle with lateral cyclic and pedals, whilst monitoring N_R			
Manoeuvre description	The EP shall perform the confined area take-off procedure as described in the AW109S Trekker Rotorcraft Flight Manual (RFM). Starting from a stabilised hover 5 ft above the ground, on a Northerly heading at the centre of the helipad, the EP will initiate a (nominal) 350 ft/min Rate of Climb (RoC) whilst maintaining sight of the helipad by translating backwards (position 1 in the figure below). The EP will continue the ascent towards the Extended Take-Off Decision Point (TDPe) while keeping the helipad in view. The aircraft will experience a single engine failure during the climb (position 2) and the EP will initiate a One-Engine-Inoperative (OEI) return to the helipad. The EP will lower collective to stop climbing and apply forward cyclic to arrest the rearwards motion and capture a descending flightpath to return to the helipad, maintaining sight of the helipad during the descent (position 3). An N_R value of 101% should be re-captured following the failure. The collective should be adjusted to cushion the touchdown (TD) as required (position 4). Rate of Descent (RoD), ground speed and track angle at TD must be within performance requirements below			
Test variations	Condition	Failure Height	Mass	Pressure Altitude
		120 ft 120 ft	3115 kg Up to MTOWkg	745 ft 9280 ft
	TDP_E		400 ft	
Test course description	Helipad, with appropriate markings situated in a confined area and in Visual Meteorological Conditions (VMC)			
Ratings scales	1. Simulation Fidelity Rating (SFR) Scale 2. Motion Fidelity Rating Scale (MFR) 3. Cooper-Harper Handling Qualities (HQR) Rating Scale			
Performance standards			Desired (*d*)	Adequate (*a*)
	Landing position from centre of helipad		± 10 ft	± 15 ft
	TD rate of descent		< 400 ft/min	< 500 ft/min
	Track angle at TD		± 5°	± 10°
	Forward ground speed at TD		3 kts	5 kts

(continued)

(continued)

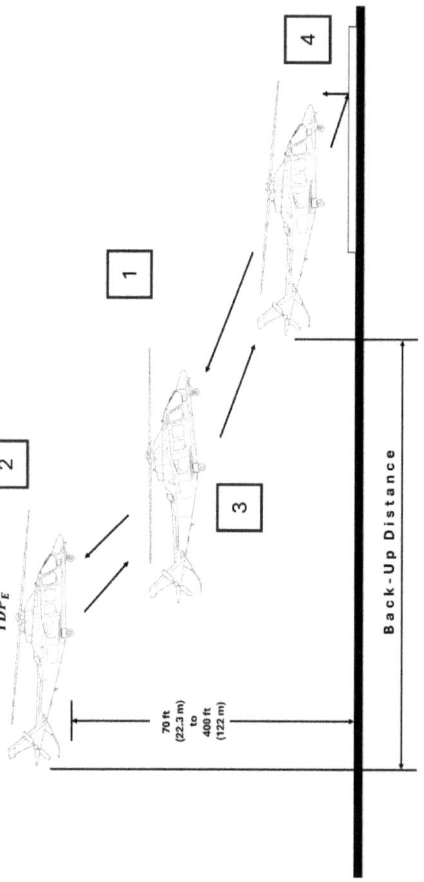

Table 13.1 RTO FTM MDA parameters

	Surge		Sway		Heave	
MDA	k_x	ω_{hpx}	k_y	ω_{hpy}	k_z	ω_{hpz}
RTO	0.5	1.0	0.5	1.0	0.7	2.8
	Roll		Pitch		Yaw	
MDA	k_ϕ	$\omega_{hp\phi}$	k_θ	$\omega_{hp\theta}$	k_ψ	$\omega_{hp\psi}$
RTO	0.4	0.4	0.35	0.4	0.87	0.6

Appendix 2: HELIFLIGHT-R Motion Drive Law Parameters Used in the CAT A RTO FTM

Simulator platform movements are determined by the MDA that scale, limit and filter the signals from the FSM to generate VeMCS commands. An example of a third-order filter used for the pitch axis MDA is given in Eq. (13.3) which scales and filters the FSM pitch acceleration, $\ddot{\theta}$, converting it into a commanded motion platform pitch acceleration, $\ddot{\theta}_s$. The parameters k_θ and $\omega_{hp\theta}$ are the pitch high pass (hp) filter gain and break-frequency coefficients, respectively which are 'tuned' for a given FTM. Similar filters are used in other rotational (ϕ, θ) and the translational axes (x, y, z). Table 13.1 lists the MDA parameters used in the CAT A RTO FTM.

$$\frac{\ddot{\theta}_s}{\ddot{\theta}}(s) = k_\theta HP_\theta(s) = k_\theta\left(\frac{s^2}{s^2+2\zeta_{hp\theta}\omega_{hp\theta}s+\omega_{hp\theta}^2}\right)\left(\frac{s}{s+\omega_{b\theta}}\right) \tag{13.3}$$

References

1. EASA (2018) Certification specifications and acceptable means of compliance for small rotorcraft CS-27/Amendment 6. EASA
2. EASA (2018) Certification specifications and acceptable means of compliance for large rotorcraft CS-29/Amendment 5. EASA
3. anon (2000) Aeronautical Design Standard Performance Specification: Handling Qualities Requirements for Military Rotorcraft (ADS-33E). U.S. Army Aviation Engineering Directorate, Redstone, Alabama
4. Federal Aviation Administration (2008) AC 29-2C—Certification of transport category rotorcraft. FAA
5. Cooper GE, Harper RP (1969) The use of pilot ratings in the evaluation of aircraft handling qualities. NASA TN D-5153
6. Leonardo S.p.A (nd) A109S Trekker—RFM Document 109G0040A034—Supplement 4, CAT Operations General
7. DuVal RW, He C (2018) Validation of the FLIGHTLAB virtual engineering toolset. Aeronaut J 122(1250):519–555. https://doi.org/10.1017/aer.2018.12
8. EASA (2012).Certification specifications for helicopter flight simulation training devices CS-FSTD(H). EASA
9. Serr C, Hamm, J., Toulmay, F., Polz, G., Langer, H. J., Simoni, M., Bonetti, M., Russo, A., Vozella, A., Young, C., Stevens, J., Desopper, A., & Papillier, P. (1999). Improved Methodology

for Take-Off and Landing Operational Procedures: The Respect Programme. 25th European Rotorcraft Forum, Rome, Italy.
10. Perfect P, Timson E, White MD, Padfield GD, Erdos R, Gubbels AW (2014) A Rating Scale for the Subjective Assessment of Simulation Fidelity. The Aeronautical Journal 118(1206):953–974. https://doi.org/10.1017/S0001924000009635
11. White MD, Perfect P, Padfield GD, Gubbels AW, Berryman AC (2013) Acceptance testing and commissioning of a flight simulator for rotorcraft simulation fidelity research. Proc Inst Mech Eng Part G J Aerosp Eng 227(4):663–686. https://doi.org/10.1177/0954410012439816
12. Podzus P, White MD, Dadswell CM, van 't Hoff S, Crijnen J (2022) Evaluation of simulator cueing fidelity for rotorcraft certification by simulation. In: 78th VFS Annual Forum & Technology Display, Fort Worth, Texas
13. Hodge SJ, Perfect P, Padfield GD, White MD (2015) Optimising the yaw motion cues available from from a short stroke hexapod motion platform. Aeronaut J 119(1211):1–21. https://doi.org/10.1017/S0001924000010228
14. Hodge SJ, Perfect P, Padfield GD, White MD (2015) Optimising the roll-sway motion cues available from a short stroke hexapod motion platform. Aeronaut J 119(1211):23–44. https://doi.org/10.1017/S000192400001023X
15. Roscoe J, White MD, Hodge SJ, Padfield GD (2022) Rotorcraft pitch-surge motion cueing requirements for a simulated offshore approach task. In: 78th VFS Annual Forum and Technology Display, Fort Worth, Texas
16. Schroeder JA (1999) Helicopter flight simulation motion platform requirements. NASA/TP-1999-208766
17. Genesys Aerosystems (2016) IDU-680 Version 8.0E, pilot operating guide and reference. Document 64-000098-080E

Chapter 14
Case Study 3: CS 27/29 Dynamic Stability Requirements

14.1 Introduction

This Chapter presents the results from the Case Study on the Dynamic Stability (DS) Applicable Certification Rule (ACR), CS 29.181, and Certification Specification (CS) 27/29 Appendix B [1, 2], to illustrate the application of the Rotorcraft Certification by Simulation (RCbS) process. Section 14.2 introduces DS as a flying quality, and describes the range of different rules for DS, with a discussion on the implications for handling qualities (HQs) and pilot workload. Section 14.3 presents the results from the case study, applying the RCbS process to the DS ACR, including the credibility of extrapolation of the findings to different flight conditions. Section 14.4 then introduces a new flight-test-manoeuvre (FTM), in the style of ADS-33E-PRF's (ADS-33's) mission-task-elements (MTEs) [3], designed to evaluate the impact of different levels of DS on pilot workload, including flight in turbulent atmospheric conditions. Results from exploratory piloted simulation trials are presented. Section 14.5 summarises the main conclusions and associated recommendations derived from this RoCS case study. Although this Chapter connects with the examples discussed in Chap. 9 of the Book, it is presented as a relatively stand-alone study.

Referring to the Category A, Visual Flight Rules (VFR, para 29.181 [2]), DS needs to be quantified in the range V_y (best climb speed) to V_{NE} (never exceed speed), while for Instrument Flight Rules (IFR, Appendix B VI), the range is between V_{MIN} (minimum IFR speed) and V_{NE}, although this upper airspeed should, in general, be further investigated. A more general starting point is illustrated in Fig. 14.1 showing how a comprehensive 'trim' test point matrix across the whole airspeed-altitude envelope might be defined; in this fictitious case, there are nine (forward) flight speeds and five density altitudes.

Note that in this hypothetical case, the high-altitude hover and high-speed cases are considered outside the flight envelope. Also, DS for airspeed cases below the lower values described above are not normally investigated in flight test certification. At each of these 43 test conditions could be added variations in flightpath angle (vertical

G.D. Padfield et al., *Rotorcraft Certification by Simulation and Analysis*, Springer Aerospace Technology, https://doi.org/10.1007/978-3-031-86398-1_14

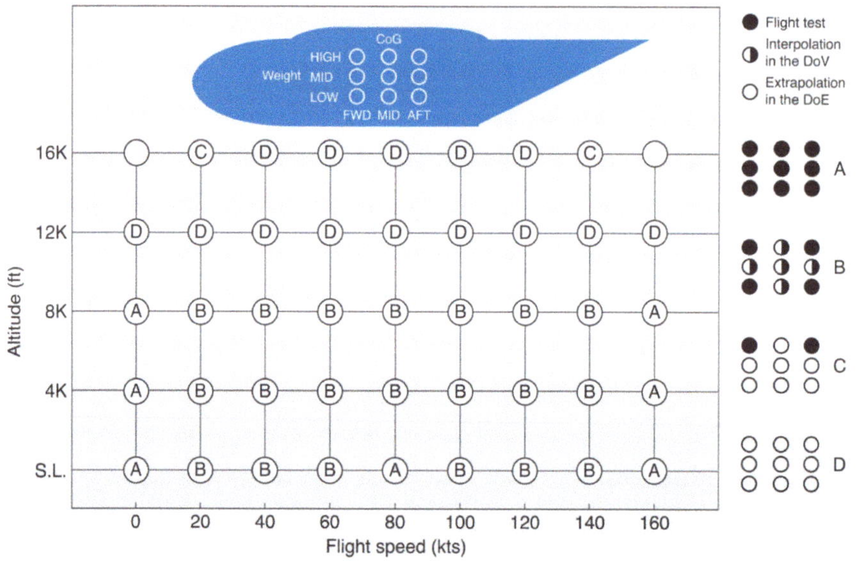

Fig. 14.1 Matrix of possible flight condition and aircraft configuration test points for DS ACR (*points identified for flight simulation testing, interpolation and extrapolation are shown in circles*)

and horizontal) and turn rate, acknowledging the changing behaviour of the aircraft in, e.g. climbing, sideslipping or turning flight. Adding a single positive and negative increment for each variable trebles the number of test conditions at each trim point, giving a possible total of 1161 test points. The flight conditions are accompanied in Fig. 14.1 by nine possible aircraft weight (low, mid, high) and longitudinal balance (fwd, mid, aft c.g.) configurations, giving an accumulated total of 10,449 points. To emphasise, this range of points is much larger than normally investigated in the CS 27/29 flight tests, summarised in Appendix 1 of this Case Study. However, quantifying the DS across the whole envelope of possible flight conditions through simulation provides a complete picture of flight behaviour related to DS, from which fidelity assessments and any underlying sources of deficiency can be investigated.

A comprehensive assessment of DS across this hypothetical forward-flight envelope might involve several hundred hours of testing and, depending on the productivity, months of elapsed time. Recalling that this is for a single ACR, it is obvious that this level of coverage is unlikely to be achieved in flight test.[1] In comparison, depending on the computing power available and the Flight Simulation Model (FSM) complexity, the dynamic and static stability results and analyses can be 'crunched' and documented within a few hours, or days at the most. Moreover, automated analyses can quantify margins with respect to certification 'boundaries' and highlight critical conditions (CPA in I2) for further investigation.

[1] Appendix 1 summarises what a typical flight-test campaign might cover, for nominal conditions; noting that for failure or degraded conditions (typically failures of the AFCS,) a different set of ACRs are relevant.

To make the assessment more realistic, Fig. 14.1 shows the test points identified as either flight tested, interpolated-simulation in the Domain of Validation (DoV) or extrapolated-simulation in the Domain of Extrapolation (DoE). With the low-medium altitude cases bookended by flight test, and the high-altitude cases bookended by the high weight, fwd and aft c.g. aircraft configurations, this arrangement gives a 62% replacement of flight test by simulation.

Of course, this form of productivity metric is only one aspect of the benefits of using simulation, but it gives a flavour of the potential. All the above are defined (and agreed on with the Certification Authority) in Phase 1 of the RCbS process, including metrics and tolerances for FSM and Flight Simulator (FS) fidelity to be used in Phase 2 and the uncertainty analysis and credibility metrics in Phase 3. The extensive activities in Phase 2 describe how the FSM/FS/FTMS developments meet the Requirements defined in Phase 1, including Verification and Validation (V&V), fidelity/uncertainty assessment and any physics-based model updating. Based on the successful achievements in Phase 2, Phase 3 can then focus on extrapolation, credibility assessment and certification.

In the next section, the DS case study is presented in detail, set within a brief historical context. An assessment of static stability ACRs would likely accompany those for DS, results from the former used to support the latter. In the present exercise, only DS is considered.

14.2 Dynamic Stability as a Flying Quality

The relevant CS-29 certification rule for Category A VFR approval in nominal conditions is CS-29.181 (Dynamic Stability). For light and low-passenger capacity rotorcraft, CS-27/29.171 states that for VFR flight, "*the rotorcraft must be able to be flown, without undue pilot fatigue or strain, in any normal manoeuvre for a period of time as long as that expected in normal operation.*" The descriptors 'undue fatigue' and 'normal manoeuvre' are not elaborated on in the specifications. For IFR flight, as presented in CS 27/29 Appendix B, the requirements are quantified in terms of how much the motion must damp within various oscillation cycles. So, for example, for single-pilot IFR, CS-27 requires that, "*any oscillation having a period of less than 5 s must damp to ½ amplitude in not more than one cycle.*" This requirement also applies to CS-29 IFR operations. Such 'flying qualities' requirements can be shown on a chart of oscillation frequency vs. damping. Figure 14.2 shows the chart with multiple boundary lines for DS that can be related to the various flight 'modes', e.g. pitch-heave short-period, phugoid or lateral-directional Dutch-roll.

Figure 14.2 also includes the ADS-33E HQ Level boundaries for minimum acceptable flying qualities for lateral-directional oscillations (LDO) in forward flight (> 45 kts), for the so-called 'all other MTEs' category [3]. Not shown on the figure, for tracking and target-acquisition MTEs, a minimum value of damping ratio, or relative damping, $\zeta = 0.35$ is required for Level 1 flying qualities, to ensure that a high level of precision can be achieved with minimal pilot compensation. The

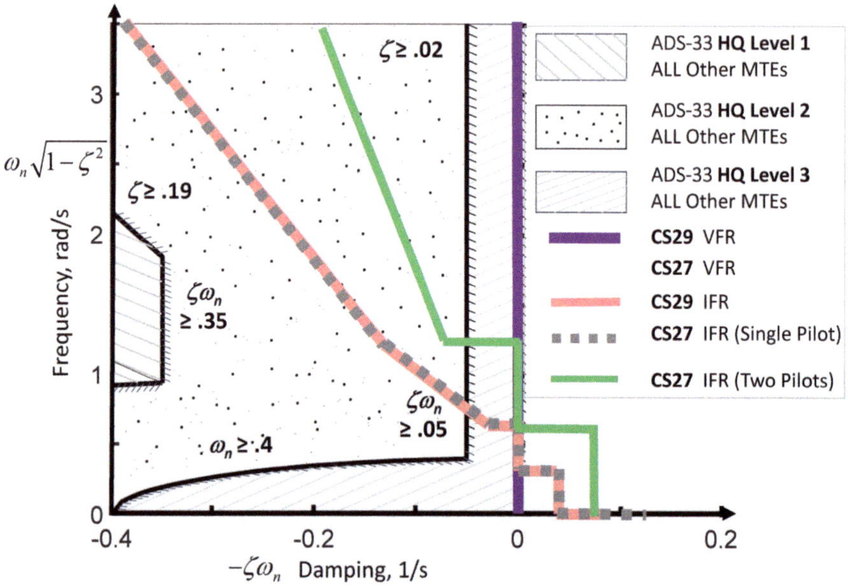

Fig. 14.2 Boundaries of acceptable DS on a chart of frequency (rad/s) versus damping (1/s); note there are no DS requirements for CS-27 helicopters approved for flight in VFR

Level 1–2 boundary for the all-other-MTEs category sits at a relative damping of about 0.19, above a frequency of about 1.8 rad/s. The ADS-33 boundaries at lower frequencies are more complex as shown in the figure, partly to harmonise with other flying qualities requirements and metrics. To achieve the Level 1 standards, ADS-33 also sets metrics for the ratio and phase of roll to sideslip/yaw in the LDO. The civil certification specifications make no reference to HQ Levels. For each category, there is a single boundary discriminating acceptable from unacceptable behaviour.

Describing the boundaries on Fig. 14.2 in more detail, flight dynamics modelling characterises each mode with an associated eigenvalue (λ) that can be derived from the linearised form of the FSM,

$$\lambda = \mu \pm i\omega \tag{14.1}$$

where the real part, μ is the mode damping and the imaginary part, ω is the mode frequency. The exponential way that such modes grow or decay, to double or half amplitude, is inversely proportional to the magnitude of the damping, i.e.

$$t_D \text{ or } t_{1/2} = 0.69/|\mu| \tag{14.2}$$

The relative damping, or damping ratio, ζ, quantifies how many oscillation cycles a mode grows or decays to double or half amplitude ($N_{1/2}$), e.g.,

$$N_{1/2} = \frac{0.69}{2\pi} \frac{\sqrt{1-\zeta^2}}{|\zeta|} \tag{14.3}$$

The mode damping and frequency are related to the relative damping (ζ) and the (undamped) natural frequency (ω_n) through the equations,

$$\mu = -\zeta \omega_n \tag{14.4}$$

$$\omega = \omega_n \sqrt{1-\zeta^2} \tag{14.5}$$

The diagonal boundaries in Fig. 14.2 are lines of constant damping ratio, reflecting that the higher the mode frequency, the larger must be the damping for acceptable flying qualities. In piloting terms, the higher the frequency of disturbed motions, the quicker pilots must be able to react to suppress deviations in flight path or attitude. So, the CS-27 single-pilot IFR line referred to earlier corresponds to a ζ of about 0.11, or 11% of critical damping. The two-pilot (green) line corresponds to a lower ζ value of about 0.055. From Fig. 14.2, at low frequency, some instability is allowed; e.g., for single pilot IFR (grey-dashed line):

(a) *Any oscillation having a period of 20 s or more may not achieve double amplitude in less than 20 s.*
(b) *Any aperiodic response may not achieve double amplitude in less than 6secs.*

The relative damping standard format was first applied in the 1950s for the military helicopter standard, MIL-H-8501 (updated in 1961, Ref. [4]) and has been adopted with minimal modification in both US and European civil certification specifications. They also feature in the fixed-wing flying qualities specification [5]. The higher the mode frequencies, the more likely pilots will need to enter 'into-the-loop' to suppress unwanted roll-yaw oscillations, particularly when manoeuvring in turbulent conditions. The ratio and phase of roll to yaw oscillations in the LDO determine the complexity of control compensation required for the suppression. In Ref. [6], the historical context to this form of flying qualities standard is described in more detail and results from FSM fidelity assessment presented based on flight testing with the National Research Council of Canada's Bell 412 and Liverpool's HELIFLIGHT-R FS, featuring the FLIGHTLAB F-B412 FSM. In the Ref. [6] research, the LDO stability and response fidelity metrics showed good matches between flight and simulation following a model-updating process involving enhanced interference modelling.

While the civil standards in Fig. 14.2 are applicable to any oscillation, this Case Study also focuses on the LDO. Without stability augmentation, the LDO frequency and damping, for mid-high speed flight conditions (between V_y and V_{NE} or V_{NEI} (IFR never exceed speed), see Appendix 1), typically lie in the middle of the Fig. 14.2 chart, with frequencies below 2.5 rad/s. While it is recognised that the unstabilised, bare-airframe, configuration might not be the most commonly found (particularly CS-29 certified) aircraft in operational service, it is featured in this Case Study to draw out the physics-based nature of the LDO characteristics; these can often be 'hidden' by the effects of the Automatic Flight Control System (AFCS).

The means to establish compliance with the DS requirements of CS-27/29 involve the pilot setting up the appropriate trim condition, applying a pedal control input (e.g., doublet) and allowing the roll-yaw-sideslip oscillations to respond freely for several cycles. The frequency and damping can then be computed from the time responses. The ADS-33 test guide [7] describes how this can be achieved in detail and notes that "*if the oscillation is nonlinear with amplitude, the requirement should apply to each cycle of oscillation.*" The civil standards do not specifically address this point. However, because of the nature of fuselage/empennage force and moment variations with sideslip, such nonlinear behaviour is likely to be a normal situation. In the present case study, the variations with amplitude are contained within uncertainty analyses, as described later in this Case Study.

The absence of supporting data for the CS boundaries in Fig. 14.2 might raise questions about their veracity. However, the ADS-33 Background Information and User's Guide, Ref. [8], also admits to little supporting rotorcraft-specific data for the HQ level boundaries. More recent simulator assessments of LDO stability reported in Refs. [9, 10] do suggest a broad range of Level 2 HQ performance across a wide relative damping range at an LDO frequency of 2.5 rad/s. The results are not conclusive and the simulations conducted in these investigations did not feature turbulence. So, while the RCbS guidance set out in this Book does not challenge the CS boundaries, the RoCS team are recommending pilot-in-the-loop testing as part of the process for DS certification. The results from preliminary simulation tests with a new FTM are presented in Sect. 14.4.

Rotorcraft fitted with a stability augmentation system (SAS) typically feature LDO characteristics well to the left of the boundary lines on the chart. So-called bare-airframe LDO characteristics are more likely to lie in the regions around the boundaries, and are the focus in this Case Study.

14.3 The LDO in the DoV and DoE

Test Aircraft and Flight Simulation Environment. The RoCS project was provided with flight test data and a FLIGHTLAB FSM of the AW109 Trekker aircraft (F-AW109) with which to exercise aspects of the RCbS process. Bare-airframe[2] flight data for trim, stability and response assessment were provided to the RoCS team, by Leonardo Helicopters, for a range of test conditions, prior to any FSM analysis. Note that in the formal RCbS process, the flight test data would be gathered in Phase 2, in conjunction with the development of the FSM and FS, and following the FTMS development (incl. V&V). Correlations between test and simulation ranged across the

[2] In the RCbS process, fidelity assessment for the bare-airframe configuration is advocated to ensure that the physical sources of any fidelity deficiencies can be more clearly determined and, if necessary, corrected.

quality spectrum. A comprehensive application of the RCbS process would include behavioural fidelity and credibility assessments for the selected ACRs, including any required FSM/FS updating and uncertainty quantification. The RoCS project resources allowed only limited coverage of these aspects, adequate to illustrate the process but not always adequate to establish sufficient fidelity or credibility.

The component-content of the nonlinear F-AW109 was illustrated previously in Fig. 6.2. Some key features are the blade-element main rotor with nonlinear lift/drag variations with incidence and Mach number, finite-state dynamic inflow, wake interference and nonlinear fuselage and empennage aerodynamic data derived from wind tunnel tests. The F-AW109 has more than 60 states, including from main rotor flap and lag motions, dynamic inflow, engine and rotorspeed dynamics and control actuators, in addition to those from the six degrees of freedom (DoF) fuselage motion.

In the following, the flight test condition—airspeed 120 kts, level at 3000 ft density altitude—features as the reference point in the DoV from which DS extrapolations are carried out. The weight and balance parameters correspond to a light-weight, aft c.g. configuration. LDO stability characteristics for all cases described in this Chapter are summarised in Appendix 2.

DS Characteristics at the (DoV) Reference Point. Figure 14.3 shows a comparison of the (bare-airframe) LDO location from simulation (predicted) vs. flight test (FT, estimated) at the reference condition. Both points are derived from the yaw rate response to a pedal doublet control input, following CS-27/29 and ADS-33 methodologies [7]. The uncertainty boxes are derived from the computation of the frequency and damping using different time periods in the response (and input amplitudes in the case of the F-AW109 simulation). As noted above, nonlinearities are one source of such variability. The flight and simulation points lie either side of the CS-27/29 IFR boundary but have significant damping margin compared with the CS-29 VFR boundary; recall that CS-27 does not quantify frequency-damping for VFR flight. For reference, the predicted (F-AW109) LDO with AFCS engaged,[3] at this flight condition, is also shown on the chart, almost reaching the ADS-33 Level 1–2 boundary ($\zeta = 0.19$) for the 'all-other-MTEs' category.

In Fig. 14.3, a 10% fidelity box is shown around the FT estimate. The F-AW109 prediction lies just outside the box. The 10% fidelity tolerance on both frequency and damping is somewhat arbitrary and would need to be justified by the applicant (and agreed with the authority) in Phase 1 of the RCbS process. The acceptable tolerance is likely to reduce as the stability 'margin' reduces. Failure to achieve the defined tolerance requires that the FSM fidelity is improved with physics-based updates.

The process of model-updating to improve fidelity received significant attention by the NATO AVT 296 working group [11]. Among the methods assessed by the group was the 'renovation technique' developed by members of the RoCS team, utilising a system identification approach to estimate the stability and control derivatives that have the greatest impact on fidelity metrics and associated 'cost functions' for the flight-simulation match errors [12]. For the example presented in Ref. [6], the LDO

[3] The AFCS/SAS configuration included damping about pitch, roll and yaw axes and pitch and roll attitude retention.

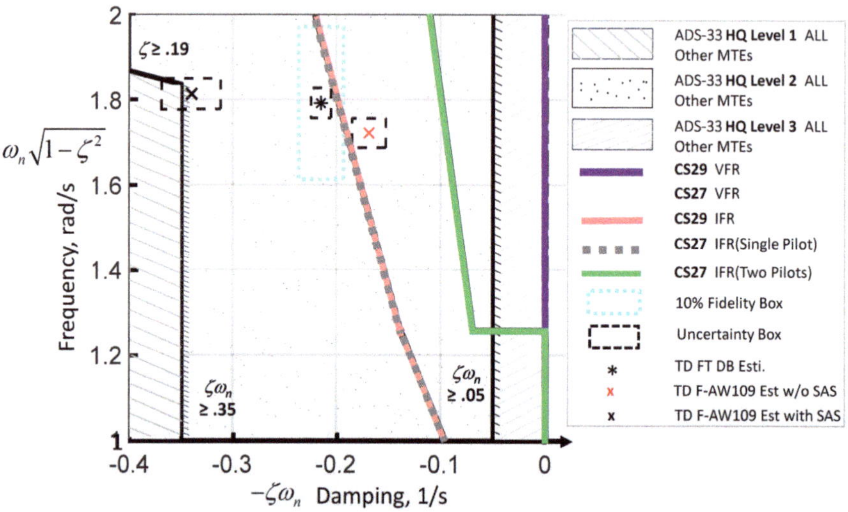

Fig. 14.3 LDO at the DoV reference point [120 kts, 3k ft; comparison of FSM (x) with flight test (∗)]; the F-AW109 SAS-on configuration is also shown on the chart (x); the frequency axis is cut-off at 1 rad/s

prediction for the Bell 412 aircraft required multiple delta derivatives, all linked with aerodynamic interference effects on the empennage and tail rotor. The renovation process then transforms the delta derivatives into auxiliary forces and moments for the relevant FSM components. This approach was also adopted in Refs. [9, 10] as part of the assessment of the ADS-33 LDO flying qualities standards.

In the present case study, a 10% delta on the yaw damping derivative, N_r, was sufficient to bring the LDO of the F-AW109 into the fidelity box, as shown in Fig. 14.4. Such a renovation can be linked to minor modifications to parameters in the interference model derived from the finite-state main-rotor inflow model. In such cases, provided the 'tuning' of such parameters is undertaken within the uncertainty ranges and relevant DoR, the updating can be considered physics-based. This is clearly a very important aspect of model-updating that applicants need to give detailed attention to.

The large margin of LDO damping for VFR flight shown on Fig. 14.4 might suggest that certification would not be an issue, but the proximity to the IFR boundary might raise concerns for such operations. In addition, as shown in Fig. 14.4, the uncertainty effectively negates the small margin. Of course, this is one of the reasons why stability augmentation is commonplace, even for CS-27 rotorcraft.

The 120 kts, 3 k ft altitude flight condition can be used to support extrapolation to a higher altitude condition, to variations in airspeed and vertical flightpath angle. As part of a 'real' application of the RCbS process, these extrapolations could, for example, be considered in the I3-P2 category (partial credit, limited extrapolation, Tables 4.2 and Fig. 4.4).

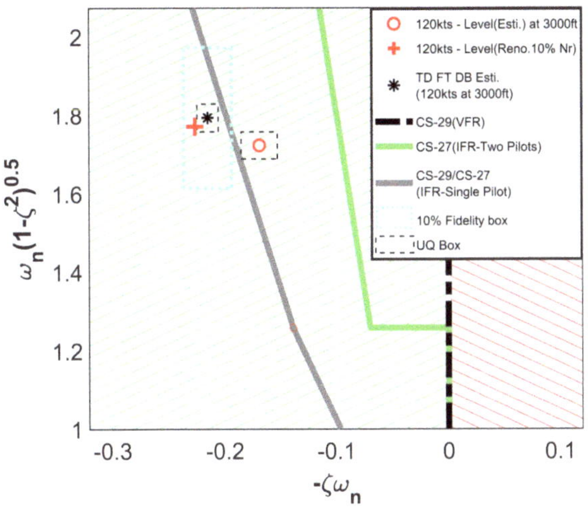

Fig. 14.4 LDO at the DoV reference point (120 kts, 3k ft) with 10% increase in yaw damping derivative N_r (+)

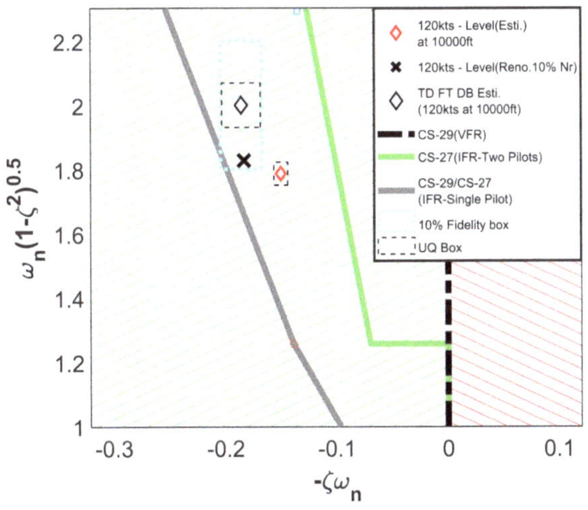

Fig. 14.5 LDO point at the 120 kts, 10 kft condition with 10% increase in yaw damping derivative N_r (x); the flight test estimate (◇) is shown for comparison

Extrapolation to a Higher Altitude Condition. The results of the application of the 10% increase in yaw damping applied at the extrapolation point, 120 kts, 10 kft, are shown in Fig. 14.5. The first thing to note is the expected reduction in LDO damping at higher altitude, a consequence of the reduced air density. As an illustration of the accuracy level of this extrapolation, the flight test point for this condition is also shown in the figure, along with corresponding uncertainty and fidelity boxes. Although the same damping renovation has brought the LDO prediction into the 10% fidelity box (which would be unknown in a real extrapolation case of course), it appears that the frequency is also under-predicted by nearly 10%. Additional renovation, e.g. with the weathercock stability derivative N_v, might be required; the physics of both yaw derivatives are connected of course, through aerodynamic interference

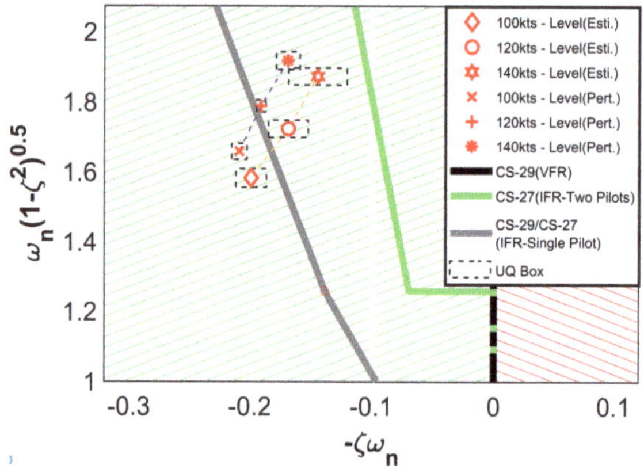

Fig. 14.6 Variation of LDO damping and frequency with airspeed, 100–140 kts; comparison of linear system eigenvalue predictions with analysis of F-AW109 yaw rate responses to pedal doublets

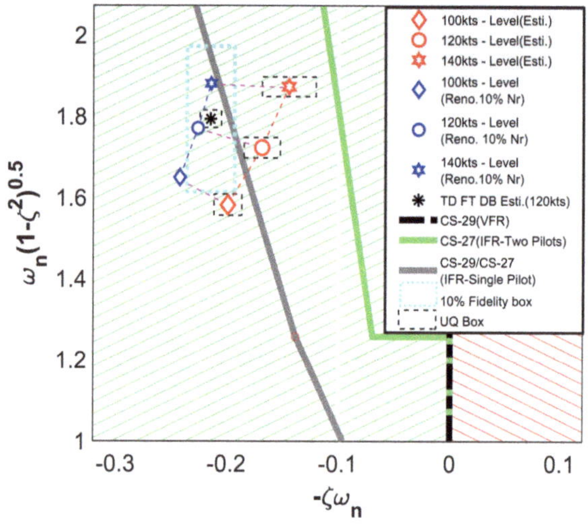

Fig. 14.7 The extrapolated LDOs with 10% renovations in yaw damping (N_r) for 100 kts, 120 kts, and 140 kts flight conditions

effects at the empennage and tail rotor. The example also highlights that extrapolations should normally be derived from more than one point in the DoV, showing trends in the variations of all aspects of the metric; in this case, frequency and damping. At an even higher altitude, say 15kft density altitude, the trends suggest that 10% renovations in frequency and damping derivatives may not provide a sufficiently high confidence in the LDO prediction, hence credibility, i.e. the extrapolation is 'likely' to be outside the fidelity tolerance, if the latter were known. Renovations in the DoE are derived from multiple DoV points and rooted in the FSM physics.

Although it might be expected that the strength of required renovations would vary continuously, even linearly, from the DoV into the DoE, new sources of fidelity deficiency can emerge that require different forms of model structure/form renovation. This example reinforces the importance of establishing the physical basis of the renovation; the physics of the renovation needs to 'keep up' with reality. The

LDO is particularly sensitive to variations in airspeed and flightpath angle, the topics examined in the next two sections.

Extrapolation to Different Airspeed Conditions. Figure 14.6 shows how the predicted LDO frequency and damping vary across the airspeed range 100–140 kts for the level flight trim condition. Recall that the reference point from the DoV is the 120 kts condition. One of the lines, labelled Level (Pert.) in the figure, corresponds to the LDO eigenvalues derived from the FSM's six degrees-of-freedom reduced-order 8×8 state matrix. The FLIGHTLAB linearisation process initially extracts the full state matrix, based on perturbing states one at a time, followed by a model-order reduction process (see, for example, Ref. [13]) to the selected number of states. This process involves physical assumptions and mathematical approximations, that both need to be justified should the linearised model predictions be used.

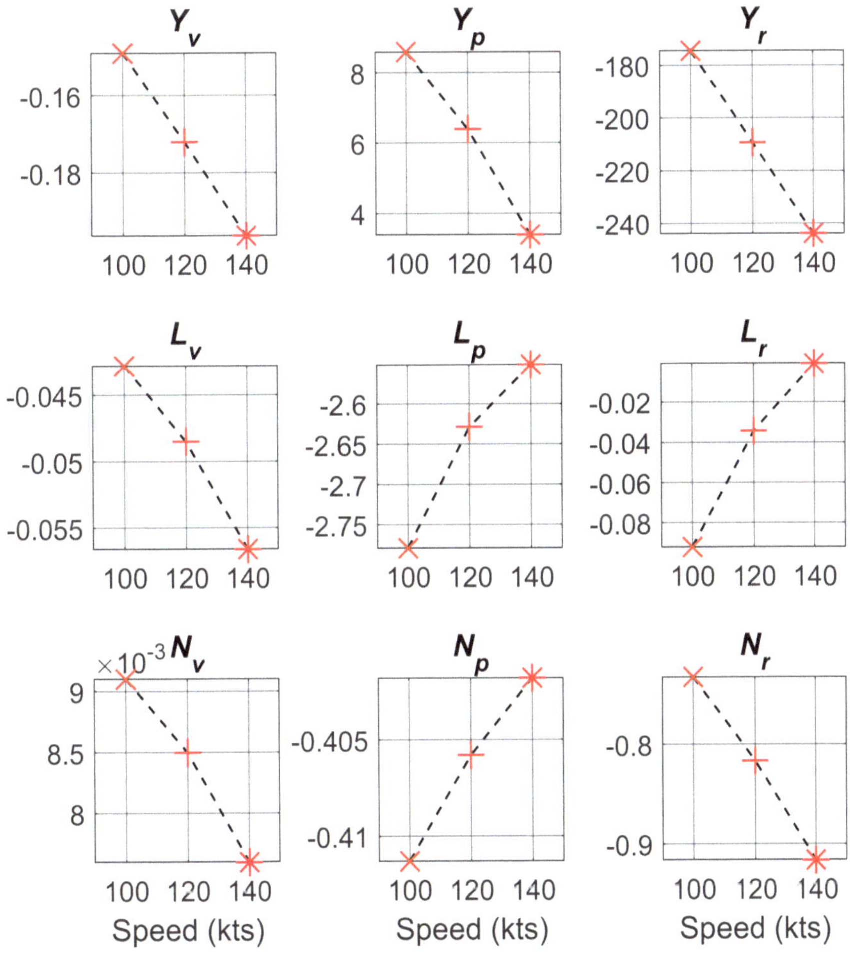

Fig. 14.8 Variations in F-AW109 lateral-directional stability derivatives as a function of airspeed

The second line in the figure, labelled Level (Est.), corresponds to the points derived from the yaw-rate oscillations of the nonlinear F-AW109 excited by a pedal doublet. In both cases the uncertainty boxes are shaped by different amplitude perturbations or control inputs. The doublet control inputs, with 2 s step duration, ranged up to 10% of maximum pedal. The large uncertainty range for the damping estimation overlaps slightly with the perturbation values for the 140 kts flight condition.

The magnitudes of the yaw rate oscillations are only one source of uncertainty relating to the estimation of the LDO characteristics. These represent quantifiable, or epistemic, uncertainties, highlighting that this metric does not represent a unique quantification of DS; a pilot may find it more, or less, difficult to suppress LDO excursions depending on the disturbance amplitude.

A key question from this analysis is what is the credibility of these, approximately linear, extrapolations to the different airspeeds? Fig. 14.7 shows the result of renovating the F-AW109 with 10% changes in damping, using the local values of N_r. With only one FT point at 120 kts, a DoV trend line cannot be shown of course, but the trend of predictions shows increasing frequency and reducing damping as airspeed increases. Can these trends be explained physically?

The variations of the stability derivatives across the airspeed range, shown in Fig. 14.8, can provide useful evidence in the search for physical explanations. The

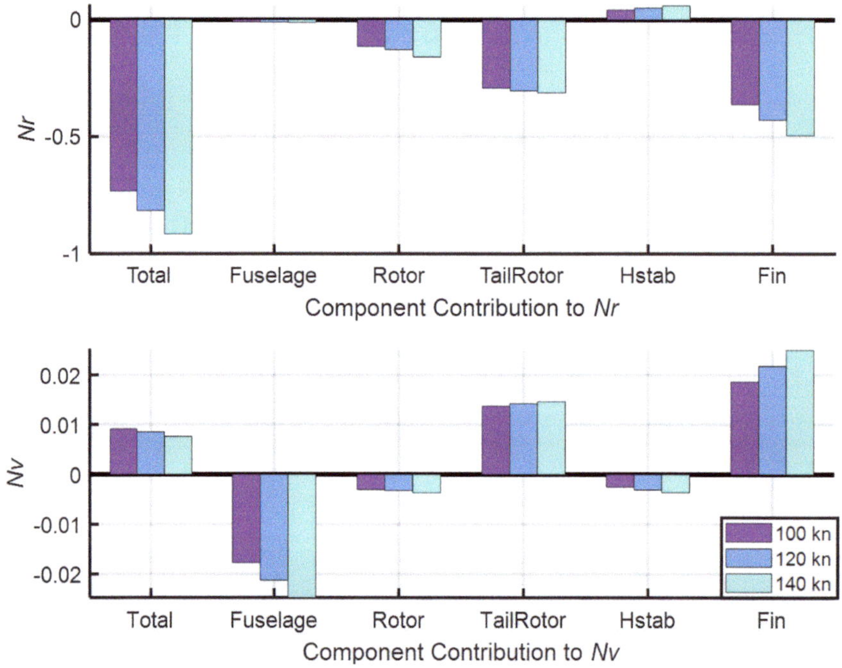

Fig. 14.9 Contributions of FSM components to the stability derivatives N_r and N_v, 100–140 kts level flight

trends for the primary roll and yaw derivatives do not suggest any strong nonlinear or discontinuous effects across this airspeed range. Further diagnostic evidence can be sought by examining derivative trends at component level.

Figure 14.9 provides an example of how fidelity assessment can be conducted at aircraft component level. The figure shows how different components, e.g. main rotor, fin, contribute to the derivatives, in this case the damping and weathercock derivatives N_r and N_v. Variations across the airspeed range are shown. Flight mechanics characteristics of interest can be explored using such figures. For example, we see that the fin contribution to the yaw derivatives shows a much larger increase with speed than the tail rotor. Both are stabilising components but the tail rotor's effectiveness saturates as forward speed increases, with the loading transformed into vibration [13].

The reducing damping evident in Fig. 14.7 has been attributed [13] to the impact of the dihedral effect, L_v, that couples with the adverse yaw N_p, when roll and yaw are out-of-phase in the LDO, to give an effective yaw damping,

$$N_{r(eff)} = N_r + N_p \frac{VL_v}{L_p^2} \tag{14.6}$$

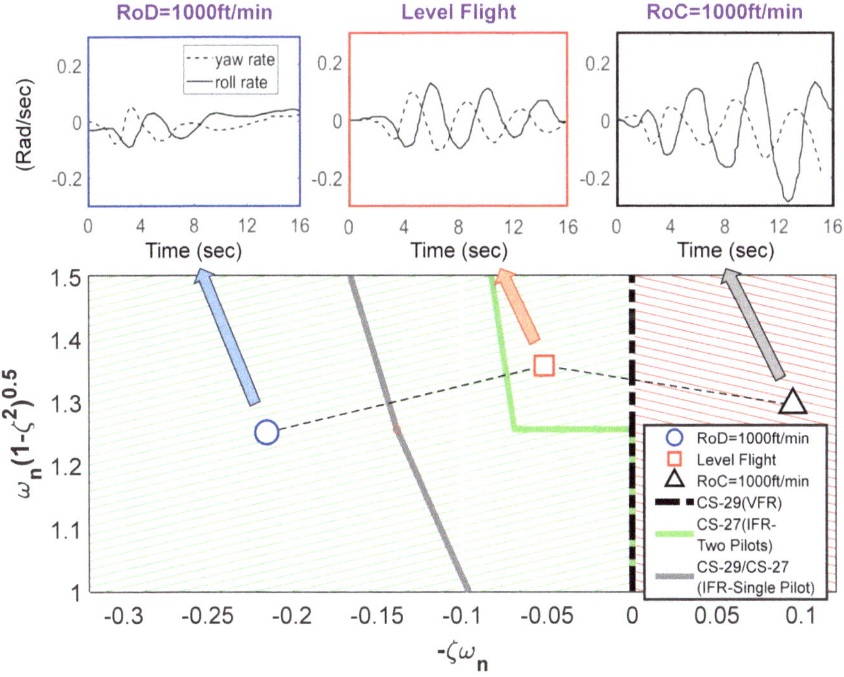

Fig. 14.10 Flight test results from RAE research Puma (SA330) showing variations of LDO stability, and roll/yaw responses to pedal doublet, with vertical flight path, 100 kts airspeed [13, 14]

The magnitude of the dihedral effect, coming largely from the main rotor and fin, increases by about 30% over the airspeed range (see Fig. 14.8). The relatively high value of adverse yaw, N_p, constant over the speed range of interest, is due to the roll-yaw coupling effects in the F-AW109 from the product-of-inertia I_{xz}, making it, effectively, proportional to the roll damping L_p over this range.

Such effective derivatives, capturing coupling effects, can provide useful physical insight but are far from the whole story. In the present case, coupling with pitch/heave dynamics also plays a part in shaping the LDO, making it very difficult to provide a complete characterisation of behaviour using such approximants.

Small perturbation, 6DoF, derivative-based models can be a valuable source of understanding the behaviour of aircraft, but their limitations should be recognised. Both the amplitude and frequency ranges over which they approximate the dynamics of the nonlinear, multi-state system, are limited and should be assessed before their credibility can be determined. Nonlinear aerodynamics, on the fuselage and empennage particularly, and couplings with higher-order modes, shape these limitations. But such degrading dynamic (LDO) stability, along with improving static and spiral mode stability, are effects that feature in both rotary and fixed-wing aircraft classical flight mechanics. One significant difference is that fixed-wing aircraft rarely fly with negative incidence (α), while this is normal for helicopters in cruise. A consequence is that a perturbation in roll (ϕ) results in an adverse sideslip $V\alpha\phi$, when α is negative [13]. This has a particularly adverse consequence in climbing flight.

Extrapolation to Different Vertical Flightpath Conditions. As an illustration of the impact of flightpath angle on LDO stability, the composite Fig. 14.10 shows results from flight tests on the Royal Aircraft Establishment's (RAE's) research Puma [14] in descent, level and climb conditions at 100 kts airspeed. The loss of stability as the flightpath angle changes from approximately 6° descent, through level to 6° climb is striking. The roll/yaw ratio varies from less than one to greater than 2, one

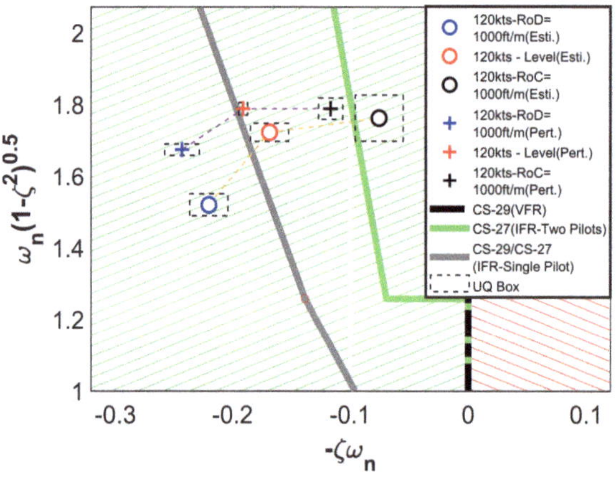

Fig. 14.11 Variation of F-AW109 LDO damping and frequency with vertical rate at 120 kts airspeed, − 1000 ft/min, level, + 1000 ft/min

consequence of the adverse sideslip effect described above. These results are for the bare-airframe Puma and contrast with the significantly increased stability provided by the stability and control augmentation system.

Figure 14.11 shows results from the F-AW109 pedal doublet analysis (Esti.) compared with linearised perturbation analysis (Pert.), again centred around the 120 kts level reference flight condition. Flight test data in climbing/descending conditions were not available to the RoCS team, so these conditions are firmly in the DoE as presented here. Based on the physical reasoning concerning the negative impact of L_v, the (50%) loss of stability in the climb, relative to the level flight case, is expected, and the magnitude of this loss is credible. The aircraft trim pitch angle/incidence varies from $-1.5°$ to $-4°$ from the descent to climb condition.

Figure 14.12 shows the impact of the 10% N_r renovations that proved successful for the DoV level flight reference case in the DoV. The model update increases the stability as expected, but before high confidence in the results can be claimed, the physical sources of the updates should be explained. The increase in the dihedral, L_v, and consequential increase in the LDO roll content are the primary sources for the damping reduction in the climb case. The results from Fig. 14.10 suggest that the HQ degradation might be expected to be larger than predicted for this aircraft. If comparative results from other types are used to support or challenge the credibility then, clearly, predictions from the comparative results are an important part of the investigation.

Credibility in Extrapolation; A Discussion. Credibility relates, on the one hand, to the level of understanding that the modeller has in the results of their predictions and, on the other, to the confidence in the level of uncertainty in both the model and test results. Uncertainty analysis goes beyond traditional modelling and simulation perspectives, which tend to focus on model-form and parameter accuracy, and rather draws on aspects such as statistical variability in design parameters and measurement

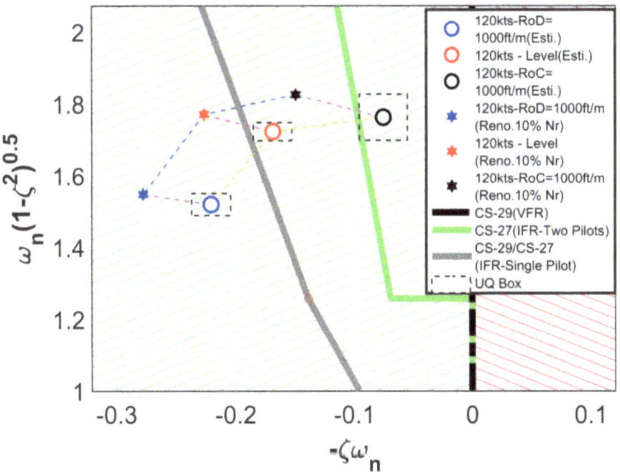

Fig. 14.12 Variation of F-AW109 LDO damping and frequency with vertical rate at 120 kts airspeed, $-$ 1000 ft/min, level, + 1000 ft/min, with 10% renovations in damping (N_r)

'process noise'. Also, the processes of solution and code verification, that are crucial to reducing uncertainty, are firmly part of RCbS Phase 2. But, outstanding uncertainties can prevail, particularly in time-constrained real-time operation of the FSM in piloted simulation. Being aware of, and able to quantify, uncertainty is recognised as a significant challenge, and the closer the model predictions are to a performance limit or stability boundary, the more important it becomes to have a measure of uncertainty to support credibility. The RCbS guidance advocates the use of the confidence ratio (CR) concept (see Chap. 4), M/U; in general terms, the smaller the margin (M) to the limit, the smaller should be the uncertainty (U) to ensure sufficient confidence in the trust of predictions. In the DS ACR, the margin relates to the level of relative damping (ζ), for the LDO. Taking the CS-27 (2-pilot) IFR boundary in Figs. 14.11 and 14.12, the renovated model prediction in the climb condition has a small positive 'stability' margin, but the uncertainty box overlaps the boundary. The CR for this case would be very small, likely spoiling the credibility of the prediction.

The LDO stability margin is only one of the metrics in the so-called predicted handling qualities (PHQs) set of ADS-33. While the minimum relative damping for the 'all-other-MTEs' category is 0.19, requirement criteria for attitude bandwidth, attitude quickness and control power, as well as multiple cross-coupling metrics must be met for an aircraft to be deemed Level 1, according to ADS-33; metric ranges also characterise Level 2 or Level 3 HQs. No such HQ metrics feature in either of CS27 or CS29, but requirements are rather expressed in qualitative terms related to 'controllability and manoeuvrability'. Within the RCbS process, the PHQ metrics are suggested as appropriate measures for the assessment of FSM fidelity, and provide expectations of results in the DoE. They can also be used as sources of evidence for HQ deficiencies that might be 'discovered' in flight test or simulation. As illustrated more fully in Chaps. 12 and 13, the RCbS process therefore recommends the design and conduct of Flight Test Manoeuvres (FTMs), in the style of ADS-33 MTEs, as part of the fidelity and credibility assessments and ultimately, if agreed with the certification authority, in the certification phase itself. Such FTMs, with their performance and workload standards, and emphasis on identifying HQ deficiencies, complement the off-line predictions, and, where applicable, can replace flight test. The next section describes preliminary results from the exercising of a new FTM for the DS ACR.

14.4 Piloted Simulation Assessment of the Impact of the LDO on Handling Qualities

The 45 T (45° Turn) VFR FTM is described in Table 14.1 in the typical MTE format. As described in the figure, the pilot is required to fly at 500 ft above the ground, following a northerly track at 120 kts airspeed, then make a 45° track change to re-trim along the new ground-track. The task can be flown in level, climbing or descending flight and the presence of an initial headwind and turbulence are intended to make the task more difficult, increasing the excitation of oscillatory modes like

Table 14.1 The 45° turn (45 T) flight test manoeuvre

Mission	45° Turn (45 T) in VMC
Critical HQs	LDO stability Attitude bandwidth and quickness, Cross-couplings: pitch/roll, roll/pitch
Objectives	• Assess the suitability of the (LDO) stability margins defined by CS27/29 through piloted simulation assessment • Check ability to perform flight path and speed control in a lateral flight path change manoeuvre in the presence of wind and atmospheric turbulence—in level and climbing flight • Assess utility of FS to extrapolate the level flight results to climbing flight • Assess the effect of vestibular motion cueing on task performance, control compensation and pilot perception of simulation fidelity
Manoeuvre description	The aircraft will be trimmed at a cruise airspeed *V* of 120 KIAS at a height of 500 ft above ground (3000 ft density altitude), on a nominal track angle 360, in the presence of a 20 kts headwind with 3-dimensional atmospheric turbulence. The trim bank angle should be zero. The EP will be following a line on the ground and, at a defined point in space, should manoeuvre to change heading (using approximately 30° angle of bank) to re-establish level flight following a second line on the ground oriented at 045 (right turn RT). Having stabilised on the new heading/track, the EP should announce 'stable' and maintain the flight condition for 5 s. The FTM time should be about 20–25 s. To hold the new track angle, the EP should adjust the cyclic and pedal/sideslip to maintain zero bank angle A first extrapolation case is a repeat of the above at a pressure altitude of 10,000 ft A second extrapolation case will be flown trimmed in a climb of 1000 ft/min; the initial conditions should be such that the aircraft reaches the same point in space (500 ft agl) at the start of the turn and maintains rate of climb throughout the manoeuvre

(continued)

Table 14.1 (continued)

Test course description	The manoeuvre starts on Runway 36 and there two runways oriented at ± 45° to it for the left and right turns. The width of the runways (200 ft) indicates the limit of the desired lateral track performance, and the limit of the adequate performance is indicated by pylons which are longitudinally spaced at 500 ft 		
Ratings scales	1. Cooper-Harper Handling Qualities (HQR) Rating Scale 2. Motion Fidelity Rating Scale (MFR)		
Performance standards		Desired (d)	Adequate (*a*)
	Maintain altitude δh: or, Rate of climb (1000 ft/min)	± 50 ft ± 200 ft/min	± 100 ft ± 400 ft/min
	Maintain airspeed δV	± 5 kts	± 10 kts
	Maintain lateral track after line capture	± 100 ft	± 200 ft
	Bank angle during tracking the ± 45° runway, post stable 'call'	± 5°	± 7.5°

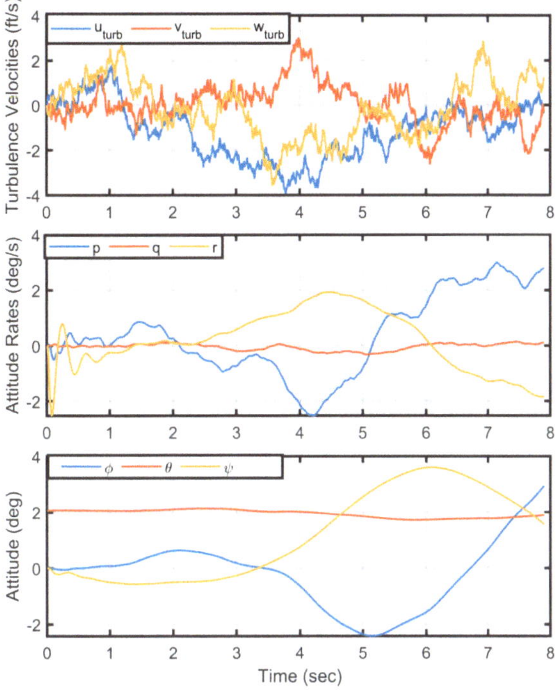

Fig. 14.13 Wind/Earth-axes turbulence components and aircraft responses

the LDO and highlighting any handling deficiencies (e.g. insufficient stability) to the pilot. The turbulence model was based on the von-Karman power-spectral-density approach [15], implemented in the form described in Ref. [16] (p. 678).

Figure 14.13 shows an 8 s slice of the turbulence components; u_{turb} in the initial headwind (north) direction, v_{turb} from west, and w_{turb} from below, and aircraft attitude responses. The aircraft angular rates, that stimulate the inner-ear vestibular (semi-circular canals) sensors, contain components of the higher frequency content, but these are mostly filtered out in the attitude responses.

As noted in Ref. [13], the Gaussian properties of such a turbulence model do not feature the intermittent features, more characteristic of structured atmospheric disturbances common in low-level helicopter flight. The statistical-discrete-gust method is proposed in Ref. [13] for such cases, and further investigations with this approach are recommended for application within the RCbS process.

Preliminary Test Results. Two test pilots participated in the workup trials, to support the 'tuning' of the motion drive laws and turbulence model parameters and refining the content of the visual database to ensure pilots could judge task performance in the VFR conditions. Three different test pilots then participated in the RCbS exercise trial with the HELIFLIGHT-R facility [17] at Liverpool in July 2023. The HQR results from the work-up and exercise trials are presented in Fig. 14.14.

During work-up, and with the initial default motion drive algorithm settings, Pilot C experienced adverse (vestibular) motion cues, particularly abrupt proverse yaw cues during the turn-in and turn-out manoeuvre phases, returning an HQR 7 for his inability to maintain the 'adequate' roll angle standards (± 7.5°) during the tracking

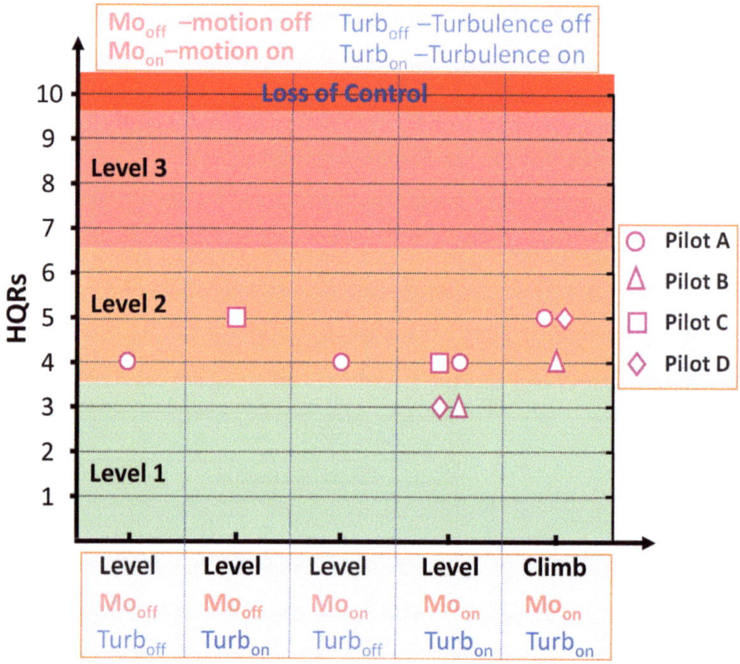

Fig. 14.14 Handling Qualities Ratings from the evaluation pilots flying the 45 T at 120 kts

phase. With the tuned roll-yaw gains and washout parameters, Pilot C achieved the desired standards but with moderate compensation in roll control during tracking (HQR 4). Pilot C found the vestibular cues improved his ability to perform the 45 T compared with the no-motion case (HQR 5).

During the RoCS exercise trial, both Pilot B and Pilot D rated the reference case as Level 1 (HQR 3), down-rating the climb case to Level 2 (HQR 4 and 5 respectively); Pilot D had difficulties holding the climb rate and maintaining the roll angle in the presence of the weakly damped LDO. Pilot A noted that both vertical and horizontal flightpath control were achieved through cyclic with minimal activity on the pedals and collective. In contrast, Pilot B used all four controls continuously throughout the FTM.

Figures 14.15 and 14.16 provide 'snapshots' of the kind of results obtained during the trial; here pilot B is flying the rated-runs at the reference condition (HQR 3) and climb condition (HQR 4).

Figure 14.15 shows the ground track through the FTM and task performance parameters during the 5 s tracking phase. The LDO, with a period of around 3.5 s, is clearly present in both attitude and sideslip variations (Fig. 14.16), but the pilot is unable to suppress this oscillatory motion fully. The desired performance is achieved in both cases but with increased compensation in the climb case leading to the HQR 4. The lateral cyclic, used for adjusting horizontal flight path and roll angle, features a moderate frequency content ($\approx$ 4 rad/s) superimposed on the lower frequency manoeuvre demands, throughout the FTM. The pilot is applying even higher frequency longitudinal cyclic movements, presumably partly in response to the vertical component of turbulence, to maintain the desired airspeed and height/height rate.

A parameter that has proven useful for quantifying control compensation (for handling qualities analysis) and adaptation (for simulator fidelity analysis), is the 'attack' activity rate (AR), based on the control attack metric (A_η), the ratio of peak rate of control deflection, $\dot{\eta}_{pk}$, to amplitude change Δ_η [13].

$$A_\eta = \frac{\dot{\eta}_{pk}}{\eta} \quad (14.7)$$

The number of attacks/sec is then an activity metric as described more fully in Refs. [18, 19]. Figure 14.17 illustrates AR for pilot B's use of all four controls, derived from 5 s windows throughout the FTM. While general rules for relating the AR with HQRs are yet to be established, the cyclic peak AR values between 1 and 2 are consistent with Level 2 ratings documented for several MTEs in Ref. [20].

For the 45 T FTM, the primary control is identified as the lateral cyclic (X_A), even though the longitudinal cyclic (X_B) features the largest peak values during the entry and exit from the 45° turn (Fig. 14.17); rising above 2/s in the latter phase. Combining the AR values from all controls, weighted using relative attack numbers per control [18], gives the integrated (peak) AR metric as shown in Fig. 14.18, plotted

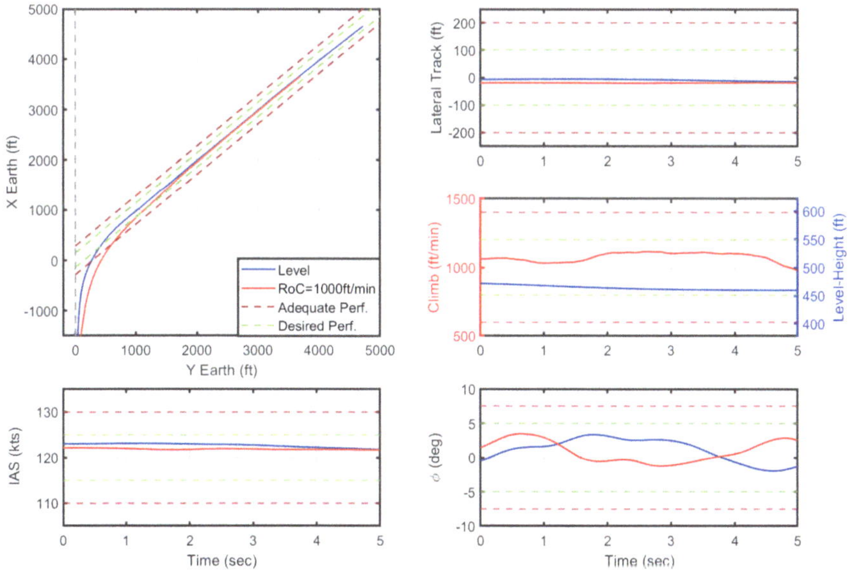

Fig. 14.15 Task performance for pilot B flying the 45 T; 120 kts reference condition/configuration in level (HQR3) and climbing (HQR4) flight

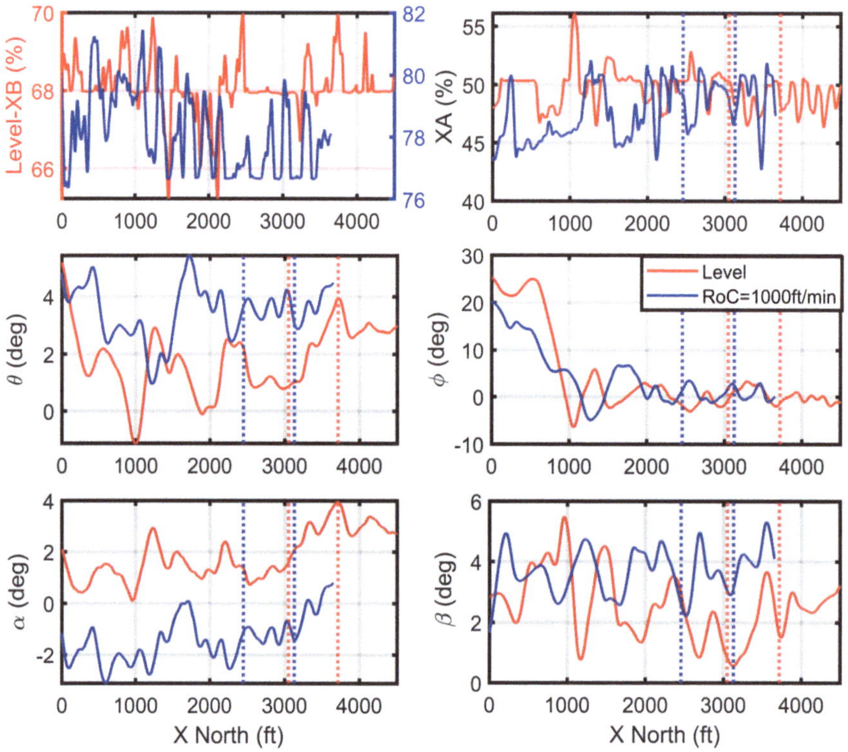

Fig. 14.16 Control activity and attitude/incidence variations for pilot B flying the 45 T; 120 kts reference condition/configuration in level and climbing flight

on the HQR chart; the individual AR values for the four controls are also shown, again noting the peak X_B attack > 2/s.

The weighted AR increases from about 1.4/s to 1.7/s as the reported level of compensation increases from minimal (HQR 3) to moderate (HQR 4).

In Phase 2 of the RCbS process, it is recommended that control compensation/adaptation metrics, such as the attack activity rate, are used in concert with the Simulation Fidelity Rating [21] scale as part of the predictive/perceptual fidelity assessment.

To emphasise, the 45 T was designed to be part of the fidelity assessment of the flight simulator in Phase 2 of the RCbS process. However, without flight test data, such a fidelity assessment was not possible for this case study.

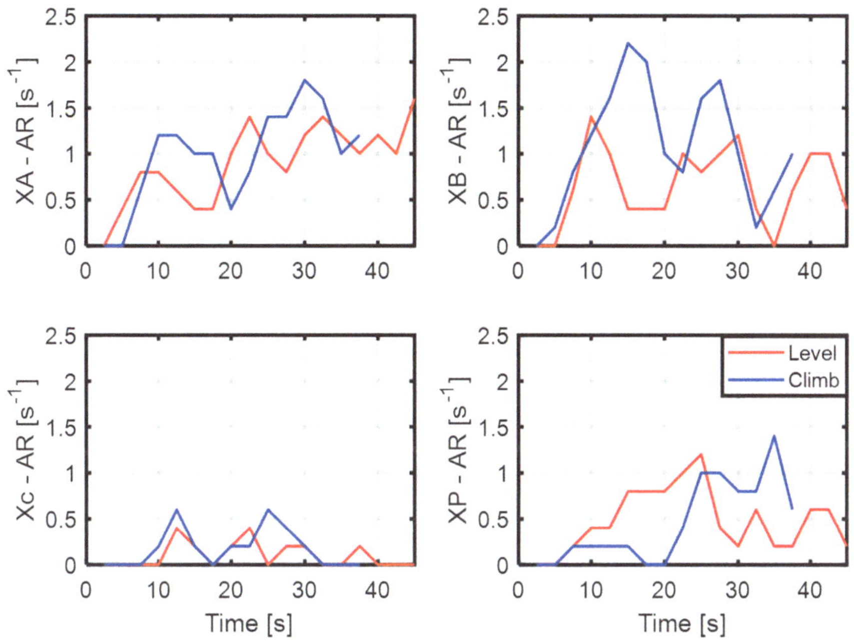

Fig. 14.17 Attack activity rate for all controls throughout the 45 T FTM; pilot B (lateral cyclic X_A, longitudinal cyclic X_B, collective X_C, pedal X_P)

14.5 Concluding Remarks and Recommendations from the Dynamic Stability Case Study

This Chapter has reported on an exercise of the RCbS process and presented results from the Case Study on the DS ACR, as expressed in the EASA certification specifications CS27 and CS29. The standards have endured for decades, and while the activity is not intended to challenge the veracity of the standards themselves, it was inevitable that questions would emerge during the study and relevant aspects have been discussed accordingly.

While it is acknowledged that most CS27/29 aircraft feature some form of stability augmentation, the RCbS process has been exercised here on the bare-airframe (AW109 Trekker) configuration, to draw out the physical sources of flight behaviour, a strongly recommended element of the process. The RoCS team were provided with DS flight test data for the bare airframe configuration of the AW109 Trekker.

A general observation from this Case Study is that DS can vary considerably from 'normal' straight and level flight conditions examined in the means of compliance assessment. With 'sufficient' fidelity, simulation provides the vehicle for assessment outside this normal. When such assessment is based on extrapolation, fidelity sufficiency should first be quantified within the DoV that encompasses such conditions, e.g. climbing/descending, turning, sideslipping flight. It is therefore recommended

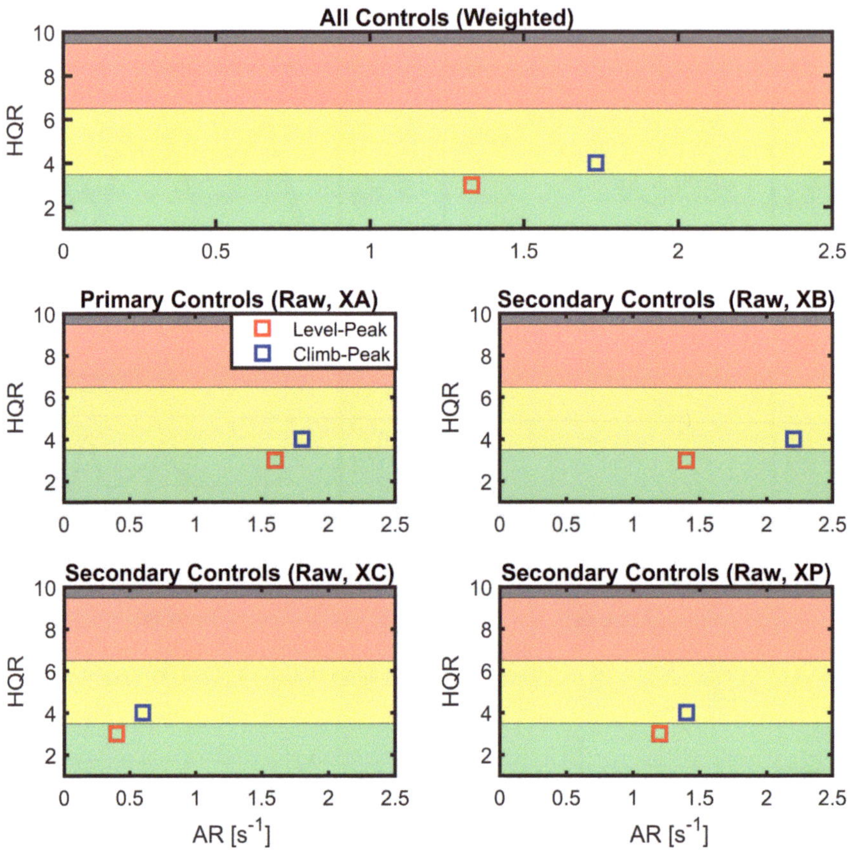

Fig. 14.18 Pilot B's peak attack activity rate versus HQR for the 45 T FTM

that such conditions be clearly defined by applicants within the DoP during Phase 1 of the RCbS process.

The results of the DS analyses have illustrated how the RCbS process can be exercised for an ACR where specific 'performance' requirements are quantified; in this case for the minimum acceptable damping ratio across the frequency range of the lateral-directional oscillatory (LDO) mode. In the Case Study, the DoV reference condition, 120kts level at 3000ft, was used to derive a FSM update that provided fidelity sufficiency for that condition, based on matching the stability metrics within 10% of flight test data. The update process was then applied to extrapolation cases; higher altitude, lower and higher airspeeds and flightpath angle variations. A limited credibility assessment of these extrapolations was explored using results from linear perturbation theory and associated stability derivatives. Such assessments are recommended as part of the RCbS process Phase 3 Credibility analysis.

Specific conclusions from the DS Case Study are as follows.

(i) For the DoV reference condition, renovation in the form of a 10% increase in yaw damping brought the LDO prediction into the (± 10%) fidelity sufficiency range.
(ii) While the stability margin for the aircraft at the reference condition was large for CS27/29 VFR operations and CS27 (IFR 2 pilots), it was much smaller for CS27/29 IFR single pilot operations, with the uncertainty effectively equalling the margin ($M/U1$).
(iii) Renovation of the LDO predictions at a 10,000 ft (extrapolation) point showed that the same 10% update in yaw damping again achieved sufficient fidelity. However, a further update to the physics behind the frequency prediction would likely be required to strengthen confidence in extrapolations to even higher altitudes.
(iv) The ± 17% airspeed extrapolation case showed near-linear variations in the LDO stability predictions. Credibility in these predictions was reinforced by considering the physics highlighted within the stability derivatives. It is emphasised that this is only one element, albeit a very important one, of the credibility analysis.
(v) Extrapolations about the reference condition to climbing and descending flight revealed much stronger nonlinearities in the predictions. The damping margin in the climb condition (1000 ft/min) reduced below the CS27 (2-pilot IFR) boundary, with a large spread in uncertainty stemming from nonlinearities in the pedal response oscillations.
(vi) A flight test manoeuvre was designed to assess the veracity of the CS DS standards and for use in the RCbS, Phase 2, FS fidelity assessment. The manoeuvre was flown by 4 pilots in the Liverpool HELIFLIGHT-R flight simulator, where the pilot-returned (largely Level 2) HQRs concurring with expectations, based on predicted (ADS-33) handling qualities levels. The results are described as preliminary as further/deeper analysis would be required to better understand the control strategies and levels of compensation used, key elements of the simulator fidelity assessment. Such an FTM can be used in RCbS Phase 2, direct comparisons with flight results then used to build the DoV for the flight simulator.

Appendix 1: Summary of Dynamic Stability Requirements in Nominal Conditions[4]

The dynamic stability requirements in the civil certification rules are defined in the following paragraphs.

---

[4] Summary of material provided to the RoCS team in a private communication from EASA. 'Nominal conditions' means all SCAS functions required for certification are operational (no failures); this includes the case where no SCAS functions are required.

	VFR	VFR Category A	IFR
CS 27	No dynamic stability requirements	No dynamic stability requirements	Appendix B VI (a) Appendix B VI (b) *See Notes (1) and (2)*
CS 29	No dynamic stability requirements	CS 29.181 *See Note (4)*	Appendix B VI *See Note (3)*

Notes:

(1) CS 27 Appendix B VI (a) applies to single pilot IFR
(2) CS 27 Appendix B VI (b) applies to dual pilot IFR
(3) CS 29 Appendix B VI doesn't distinguish between single and dual pilot but the most stringent requirements of CS 27 Appendix B VI are applied.
(4) The following main differences between CS 29.181 and Appendix B VI requirements are to be noted:

(i) CS 29.181 only applies between V_Y and V_{NE} and the requirement asks only for positive damping, while,
(ii) Appendix B VI applies between V_{MINI} and V_{NEI} and the damping is a function of the frequency of the oscillations,
(iii) With V_{MINI} and V_{NEI} being the minimum airspeed for IFR and the V_{NE} for IFR (sometimes different from the standard V_{NE}).

A typical test matrix (minimum set of points) for a CS 29 IFR approval would include at least 5 airspeeds at a minimum of 3 different altitudes, with the appropriate landing gear configuration (as shown below). The cruise points may be flown in turns if the pitch rate effect is suspected to impact dynamic stability. The tests are flown at most critical weight and cg conditions, plus permutations of weight and cg location configurations as required.

	Slow speed	Medium speed	High speed	Climb	Descent	Approach
Sea level	V_{MINI}	Cruise	V_{NEI}	V_{YI}	Descent speed	Recommended approach speed
Medium altitude (e.g.7000 ft pressure alt.)	V_{MINI}	Cruise	V_{NEI}	V_{YI}	Descent speed	Recommended approach speed
High Altitude (e.g.15,000 ft pressure alt.)	V_{MINI}	Cruise	V_{NEI}	V_{YI}	Descent speed	Recommended approach speed

Appendix 2: LDO Stability Characteristics (Damping and Frequency)

100 kts, Level, 3000 ft, Esti	– 0.1993 + 1.5806i
100 kts, Level, 10% of N_r Reno., 3000 ft	– 0.2427 + 1.6484i
120 kts, RoC = 1000 ft/min, Esti	– 0.0755 + 1.7613i
120 kts, RoC = 1000 ft/min, Pert	– 0.1168 + 1.7877i
120 kts, Level, 3000 ft, Esti	– 0.1687 + 1.7215i
120 kts, Level, 3000 ft, Pert	– 0.1912 + 1.7871i
120 kts, Level, 10,000 ft, Estil	– 0.1500 + 1.7864i
120 kts, RoD = 1000 ft/min, 3000 ft, Esti	– 0.2205 + 1.5183i
120 kts, RoD = 1000 ft/min, 3000 ft, Pert	– 0.2435 + 1.6734i
120 kts, Level, 10% of N_r Reno., 3000 ft	– 0.2264 + 1.7694i
120 kts, RoC = 1000 ft/min, 10% of N_r Reno, 3000ft	– 0.1492 + 1.8243i
120 kts, Level, 10% of N_r Reno, 3000 ft	– 0.2264 + 1.7694i
120 kts, RoD = 1000 ft/min, 10% of N_r Reno, 3000 ft	– 0.2775 + 1.5460i
120 kts, Level, 10% of N_r Reno., 10,000 ft	– 0.1831 + 1.8277i
120 kts, FT 3000 ft, Esti	– 0.2149 + 1.7921i
120 kts, FT 10,000 ft, Esti	– 0.1856 + 2.0006i
140 kts, Level, 3000 ft, Esti	– 0.1445 + 1.8712i
140 kts, Level, 10% of N_r Reno., 3000 ft	– 0.2144 + 1.8790i
Puma, 80 kts, RoC = 1000 ft/min, Esti	0.0949 + 1.2954i
Puma, 80 kts, Level Flight, Esti	– 0.0518 + 1.3572i
Puma, 80 kts, RoD = 1000 ft/min, Esti	– 0.2149 + 1.2522i

References

1. anon (2018) Certification specifications and acceptable means of compliance for small rotorcraft CS-27. EASA
2. anon (2018) Certification specifications and acceptable means of compliance for large rotorcraft CS-29. EASA
3. anon (2000) Aeronautical Design Standard Performance Specification: Handling Qualities Requirements for Military Rotorcraft (ADS- 33E). U.S. Army Aviation Engineering Directorate, Redstone, Alabama
4. anon (1961) Military specification—general requirements for helicopter flying and ground handling qualities. MIL-H-8501A
5. Moorhouse DJ, Woodcock RJ (1982) Background information and user guide for MIL-F-8785C, military specification-flying qualities of piloted airplanes. U.S. Air Force Wright Aeronautical Labs

6. Agarwal D, Lu L, Padfield GD, White MD, Cameron N (2021) Rotorcraft lateral-directional oscillations: the anatomy of a nuisance mode. J Am Helicop Soc 66(4):1–13. https://doi.org/10.4050/JAHS.66.042009
7. Blanken CL, Hoh RH, Mitchell DG, Key DL (2008) Test guide for ADS-33E PRF. Special Report AMR-AF-08-07, AMRDEC
8. Key DL, Blanken CL, Hoh RH, Mitchell DG, Aponso BL (2015) Background information and user's guide (BIUG) for handling qualities requirements for military rotorcraft. U.S. Army Research, Development, and Engineering Command, Special Report RDMR-AD-16-01
9. Cameron N, Memon WA, White MD, Padfield GD, Lu L, Agarwal D (2021) Appraisal of handling qualities standards for rotorcraft lateral-directional dynamics. AIAA SciTech Forum (virtual). https://doi.org/10.2514/6.2021-0592
10. Cameron N, White MD, Padfield GD, Lu L (2021) Appraisal of rotorcraft handling qualities requirements for lateral-directional dynamics. In: 47th European Rotorcraft Forum, Glasgow
11. Tischler MB, White MD, Padfield GD et al (2021) Rotorcraft flight simulation model fidelity improvement and assessment. NATO Report, No. STO-TR-AVT-296-UU. https://doi.org/10.14339/STO-TR-AVT-296-UU
12. Lu L, Padfield GD, White MD, Perfect P (2011) Fidelity enhancement of a rotorcraft simulation model through system identification. Aeronaut J RAeS 115(1170):453–470. https://doi.org/10.1017/S0001924000006102
13. Padfield GD (2018) Helicopter flight dynamics, 3rd ed. Wiley, London. https://doi.org/10.1002/9781119401087
14. Padfield GD (1985) Flight testing for performance and flying qualities. Helicopter Aeromechanics, AGARD LS-139
15. von Karman T (1961) Progress in the statistical theory of turbulence: classical papers in statistical theory. Interscience Publications
16. anon (1997) Flying qualities of piloted aircraft. MIL-HDBK 1797, US Department of Defense Handbook
17. White MD, Perfect P, Padfield GD, Gubbels AW, Berryman AC (2011) Acceptance testing and commissioning of a flight simulator for rotorcraft simulation fidelity research. Proc Inst Mech Eng Part G J Aerosp Eng 227(4):663–686. https://doi.org/10.1177/0954410012439816
18. Memon WA, White MD, Padfield GD, Cameron N, Lu L (2022) Helicopter handling qualities: a study in pilot control compensation. Aeronaut J RAeS 126(1295):152–186. https://doi.org/10.1017/aer.2021.87
19. Memon WA, Cameron N, White MD, Padfield GD, Lu L (2021) The development of a pilot control adaptation metric for simulation perceptual fidelity assessment. In: 47th European Rotorcraft Forum, Glasgow, Scotland
20. Perfect P, White MD, Padfield GD, Gubbels AW (2013) Rotorcraft simulation fidelity: new methods for quantification and assessment. Aeronaut J RAeS 117(1189):235–282. https://doi.org/10.1017/s0001924000007983
21. Perfect P, Timson E, White MD, Padfield GD, Erdos R (2014) A fidelity scale for the subjective assessment of simulation fidelity. Aeronaut J RAeS 118(1206):953–974. https://doi.org/10.1017/S0001924000009635

Chapter 15
Finale and Potential Routes to the Adoption of RCbS

15.1 Introduction

Here, in this final Chapter, several key conclusions are drawn out from the Book as a 'finale', along with potential routes forward, and how the first steps along these might be taken by early adopters of the RCbS process. As emphasised throughout this Book, building an RCbS team with a broad range of skills and knowledge is considered critical to successful adoption; some relevant ideas and recommendations are presented.

15.2 Follow Process

The approach to RCbS principles and practices laid out in this book is fundamentally process-oriented, built around four Phases. In Phase 0, an applicant will create a RCbS project management plan (PMP) in harmony with their Certification Plan to be reviewed for approval by the Certification Authority. The PMP will describe the contents of the three subsequent RCbS Phases and set the achievement goals, the chosen Levels in the Influence-Predictability matrix, for the use of simulation in the selected Applicable Certification Rules (ACRs). Phase 1 assembles the RCbS Requirements Specification defining metrics and tolerances for sufficient fidelity (in the trim-stability-response construct), for the maximum acceptable uncertainties and minimum confidence ratios. Phase 2 develops, verifies and validates the enabling tools—the flight simulation model, flight simulator and flight test measurement system. Phase 2 is intensely constructive and iterative, enabling an updating of the Requirements Specification to confirm the initial selections or to 'raise or lower the bar' if appropriate. Phase 3 then focuses on the credibility of the cases for certification.

G.D. Padfield et al., *Rotorcraft Certification by Simulation and Analysis*, Springer Aerospace Technology, https://doi.org/10.1007/978-3-031-86398-1_15

Following process is, of course, a familiar 'strapline' to engineering practice that facilitates quality, timeliness and efficiency.

15.3 Requirements Capture

Our book emphasises two fundamental aspects of requirements capture/building in the RCbS process. First, it should be thorough and systematic, deploying the rigour of the systems engineering discipline. Second, it should be sufficiently flexible that the iterative aspects of the RCbS process are enabled. This will be particularly important during the formative 'days' in the life of an RCbS process, when for example, metrics and tolerances set in Phase 1 are largely derived from current practices that may benefit from improvement in Phase 2. Allowable uncertainties for measurements or input parameters may prove to be unnecessarily demanding or too weak to support credibility in Phase 3. Flexibility as a property should enable requirements-management to be recognised as an ongoing activity throughout the process.

15.4 Resource RCbS

In Chap. 10, 'Resourcing the Process' was discussed and a key point was made that, "*early applications are likely to be modest in their aims, but it is strongly recommended that the long-term aspiration to achieve the I-P full credit goals define the backbone of the capability development.*" This is reinforced here because, while there may be opportunities for short term success, the 'quick-wins' as they are sometimes described, the more extensive benefits from RCbS can only be realised by building a strong foundation and comprehensive framework as described in this Book. Any route forward must be safe (risks quantified and pitfalls avoidable), reliable (well defined with uncertainties quantified), ambitious (acknowledging the challenges) and ultimately affordable (clear returns on investment). Milestones along the route need to reflect growth in capability, in harmony with success in application. The RoCS team stress these points in view of the strategic role that CbS will play during the continuing evolution of virtual engineering within the aviation industry.

Building a RCbS team with strength in depth, that can grow 'organically', will require strategic leadership from the highest organisational level. RCbS will enable a manufacturer to 'know their aircraft', in terms of its behaviour, capabilities and limitations, throughout and beyond its safe flight envelope, considerably better than ever possible through measurements at a limited number of flight conditions.

15.5 Case Studies

Three case studies have been presented, using flight test data and a FLIGHTLAB model of the AW109 Trekker provided to the RoCS team by Leonardo Helicopters. The three ACRs were exercised on several flight simulators during the RoCS project life (DLR, NLR and Leonardo Helicopters). The results presented in the Book are taken from the simulation trials on the HELIFLIGHT-R facility at the University of Liverpool, supported by an EASA test pilot/engineer team. While it was not possible to exercise the full RCbS process in the case studies, some important insights have been gained and documented in Chaps. 12–14.

The low-speed controllability and manoeuvrability ACR highlighted both fidelity strengths and deficiencies for predicting trim pedal margins at different wind azimuth conditions. A new FTM was designed, the X-wind hover (XWH), considered more representative of manoeuvring in a wind about a hover condition than the conventional pace-car tracking method. Pilots' impressions of handling qualities deficiencies in the two FTMs were quite different, with increased compensation required in the XWH FTM to maintain performance during the yaw manoeuvres. The case study stimulated extensive discussion on the merits of stylised FTMs, with their defined performance standards, for drawing out pilot workload issues. Further work is necessary before proving the XWH FTM as a more representative test of low-speed controllability and manoeuvrability for all critical azimuth conditions.

The Category A (confined-area) rejected take-off provided valuable insight for this, potentially high-risk, ACR. The use of a virtual pilot model proved useful for quantifying the 'dynamic response' predicted fidelity across the nonlinear range of flight behaviour in the low-speed envelope. The FTM could be closely matched to flight test procedures and highlighted the importance of detail in both visual and vestibular motion cueing. The case study explored extrapolation to find the maximum weight at high altitude, highlighting rotorspeed and torque control issues prior to touchdown as the limiting factors. The case study also highlighted the benefits of simulation in exploring so-called 'abuse-cases' including procedural variations beyond those defined in the rotorcraft flight manual.

The Dynamic Stability case study initially focussed on the predicted stability metrics, the frequency and damping of oscillatory motions, at typical cruise conditions. The case included an FSM update to illustrate this aspect of achieving the required tolerances on fidelity sufficiency. Extrapolation from steady level trim condition to higher airspeeds, altitudes and increase/decrease in flightpath angle, exposed the strong sensitivity of the 2D fidelity metric to flight condition. A new FTM, the 45 °turn was designed and tested to explore pilot opinion sensitivities to these flight conditions. This enabled the assessment of the impact of moderate turbulence on pilot opinion, that typically degraded handling qualities ratings from Level 1 to Level 2.

In all three FTMs, opportunities were taken to use the novel simulation fidelity rating (SFR) scale, alongside other pilot-opinion scales, to demonstrate its utility for comparing configurations with different fidelity, e.g. motion vs no motion. The de-brief dialogue surrounding these assessments supported a deeper understanding of the importance of perceptual fidelity alongside predicted fidelity as an important aspect of credibility.

15.6 What Next for Early Applications?

The Book advocates small steps in the pursuance of big goals. Some suggestions for early adopters of RCbS to pick up on are listed below.

(a) Study and understand the RCbS process set out in this Book, particularly the value of the iterative pathways between phases.
(b) Undertake a thorough assessment, a calibration and audit, of your existing FSM/FS/FTMS capabilities, in terms of both models and facilities, and human skills and experience. Calibrations could relate to the use of both interpolation and extrapolation, and draw on existing test data.
(c) Build capability around cases, by selecting ACRs that enable the full RCbS process to be exercised, albeit at reduced levels; include uncertainty quantification in this capability development.
(d) As recommended in Chap. 10, apply the RCbS process at various I-P levels, and for specific ACRs, to existing, certified, products; exercising extrapolation in such cases could be particularly valuable, pinning the corners of the DoP using certified flight test points.
(e) For any application, e.g. from (c) or (d), assess carefully the levels of uncertainty of your M&S predictions, and data from your FTMS, to provide users of these results an indication of the credibility levels.
(f) Consider the potential utility of flight test data for M&S validation purposes as a standard part of development/envelope expansion flight test preparation activities.
(g) Scope out what a 'fully-operational' RCbS team might look like; how it might fit into the Company organisational structure and how capabilities can be grown and sustained for the long term.
(h) Train your engineering team to develop and report certification-ready M&S results, with a section in the report dedicated to evidence that supports the credibility of the results.
(i) Consider how your applications might need to comply with 'industry-wide' standards and the importance of knowledge-sharing in this context. This Book has been specific in its recommendation that early-adopters are pro-active in sharing good practice throughout the community.
(j) Seek support from relevant certification authorities who also need to develop a deep understanding of the RCbS process.

MIX
Papier aus verantwortungsvollen Quellen
Paper from responsible sources
FSC® C105338

If you have any concerns about our products,
you can contact us on
ProductSafety@springernature.com

In case Publisher is established outside the EU,
the EU authorized representative is:
Springer Nature Customer Service Center GmbH
Europaplatz 3, 69115 Heidelberg, Germany

Printed by Libri Plureos GmbH
in Hamburg, Germany